LIVING IN THE GREENHOUS

LIVING IN THE GREENHOUSE

A GLOBAL WARNING

WHAT WE MUST DO, AND WHAT WILL HAPPEN IF WE DON'T

Michael Allaby

THORSONS PUBLISHING GROUP

First published 1990 by Thorsons Publishing Group

British Library Cataloguing in Publication Data

Allaby, Michael, *1933*-
Living in the greenhouse.
1. Climate. Effects on man
I. Title
551.6

ISBN 0 7225 2258 4

Published by Thorsons Publishing Group, Wellingborough, Northamptonshire NN8 2RQ, England

Typeset by Burns & Smith Ltd., Derby

Printed in Great Britain by Mackays of Chatham, Kent

10 9 8 7 6 5 4 3 2 1

Contents

INTRODUCTION

The Birth of an Idea

An eminent scholar in several fields that today would be considered quite disparate, a renowned researcher, and an enthusiastic popularizer and lecturer, John Tyndall was the very model of a true Victorian scientist. Born in Ireland in 1820 he trained first as a surveyor and engineer but the profession seems not to have suited him. At the age of twenty-eight he enrolled at the University of Marburg in Germany and was awarded a doctorate after two years. At thirty-two he was elected a Fellow of the Royal Society and at thirty-three he became professor of natural philosophy at the Royal Institution in London. There he met, and became a colleague and friend of, Michael Faraday. Tyndall died in 1893.

The list of his accomplishments is impressive. He verified what before had been only an observation, that putrefaction does not occur in germ-free air – so contributing very directly to our modern concept of hygiene. He studied the way light and sound waves travel through air for the benefit of those who design lighthouses and fog horns. In fact, he devoted many years to the study of light, the radiation we see, and the radiation we do not see but some of which we can feel, and its effects on many substances. He liked to demonstrate how a lens made from ice could act as a burning glass, his audiences no doubt delighting in the notion that ice could start fires. He was particularly fascinated in the way light and heat, radiation of different wavelengths, affect and are affected by the media through which they pass.

Like most scholars of his day he believed fervently that it was his duty to pass on to others the fruits of his studies and, unlike some, he had a remarkable talent for explaining difficult ideas in simple language. At a time when anything more than the most elementary education, and sometimes even that, was denied to all but the wealthy, most of the truly great men and women gave public lectures. Information, they believed, should be the property of everyone, and education, leading to enlightenment, would foster continued scientific and industrial progress, and the evolution of a prosperous, contented, gentle society. The modern idea that education is a commodity to be bought and sold would have struck them, rightly of course, as barbaric and ultimately self-defeating.

It was in one of his public lectures, delivered as part of a series in 1872 and 1873 in the United States, and published in 1885, that Tyndall included the following passage:

'The invisible heat, emitted both by dark bodies and by luminous ones, flies through space with the velocity of light, and is called radiant heat. *Now, radiant heat may be made a subtle and powerful explorer of molecular condition, and, of late years, it has given a new significance to the act of chemical combination. Take, for example, the air we breathe. It is a mixture of oxygen and nitrogen; and it behaves towards radiant heat like a vacuum, being incompetent to absorb it in any sensible degree. But permit the same two gases to unite chemically; then, without any augmentation of the quantity of matter, without altering the gaseous condition, without interfering in any way with the transparency of the gas, the act of chemical union is accompanied by an enormous diminution of its* diathermancy, *or perviousness to radiant heat.'*

In these words, more than a hundred years ago, John Tyndall described the way oxides of nitrogen absorb heat. The gases he mentions are what we would now call 'greenhouse gases' and they are of great concern to us all. Tyndall also found the same property in water vapour – another 'greenhouse gas'. There is, they say, nothing new under the Sun. That may or may not be a universal truth, but the words of John Tyndall prove that the idea that some gases are transparent to heat and others absorb it – the basis of the 'greenhouse effect' – has a long history. The implications of the discovery were obvious. If common atmospheric gases respond differently to radiation of different wavelengths, this must affect the climate. If heat from the Sun passes through the air, warms the ground and oceans, and the warmth they radiate as a result passes directly through a transparent atmosphere and into space, one kind of climate will result. If some of that warmth is retained in the atmosphere, on the other hand, another, warmer, climate will result.

Our story moves from Tyndall to Svante August Arrhenius (1859–1927), the Swedish scientist and Nobel prize-winner who is regarded as one of the founders of the discipline called physical chemistry. Like others of his time he refused to be constrained within the bounds of any single branch of science and his interests were wide. They involved him in studies of the origin of galaxies and stars and of the spread of life – which he suggested was disseminated through space by spores, a theory known as 'panspermia'. He considered the possibility that Mars might be habitable.

Arrhenius was a sociable, congenial man who made many friends and he never took himself too seriously. One of his anecdotes may serve as a gentle warning to those who seize on the latest scientific theory as though it were a statement of eternal truth. On one occasion he presented his professor with an entirely new theory he had devised. The professor accepted it, and politely dismissed him. Arrhenius said: 'He explained later that he knew very well that there are so many different theories formed, and that they are almost all certain to be wrong, for after a short time they disappeared; and therefore, by using the statistical manner of forming his ideas, he concluded that my theory also would not exist long'.

In about 1896, pursuing what until then had been rather general ideas about the way the chemical composition of the atmosphere affects the radiation passing through it, Arrhenius calculated that a doubling of the concentration of carbon dioxide in the air would cause temperatures to increase by about 5°C (9°F). This may be the first direct reference to what today we call the 'greenhouse effect'. Modern computer predictions agree quite closely with this calculation, although it is only fairly recently that the precise mathematical relationship between carbon dioxide concentration and temperature has been discovered, thus allowing scientists to calculate the effect of any change in the atmospheric concentration of the gas.

You might think that this relationship summarizes the greenhouse effect succinctly, indeed that it says it all. Increase the amount of carbon dioxide in the air and the world will grow warmer, decrease it and the world will grow cooler, and at present it is increasing. If that were all there is to it, this book could end here, but the real world is not so simple.

For one thing, it resists change: the resolution of environmental problems - and the greenhouse effect is regarded by most scientists as the most serious environmental problem we face - is not helped by oversimplification. Instant diagnoses accompanied by instant solutions are often misleading, sometimes completely wrong. Even when the diagnosis is accurate the apparently obvious remedy may be ineffectual or, in extreme cases, may exacerbate the symptoms.

In this book I try to explain what the greenhouse effect is and what it may mean; to do that I must probe deeply into the past. This may seem strange, since our fears that the climate may change are directed towards the future, but any serious study of the future always begins in the past. We cannot see where we are going unless we have a clear idea of how we arrived at our present position and this is as true of the atmosphere and climate as it is of anything else.

I begin by describing the atmosphere, which is much more than a diffuse mass of inert, passive gas enveloping the earth. It is dynamic, always moving, always changing - and over the course of the history of our planet it has evolved. The reasons for and consequences of the changes it has undergone lie at the root of modern greenhouse theories.

Then I move further into the past, to the birth and subsequent development of the Sun: our star also has a history, linked intimately with our own, and we need to know something of that history if we are to understand what has happened, and what will happen in the future to the solar radiation that warms and illuminates the Earth. The interaction of the two produces the general circulation of the atmosphere and, from that, the world's climates. These, too, have a history and one that may seem more familiar, for it is not long since our ancestors experienced the major climatic change that marked the end of an ice age. That change involved a warming of the greenhouse type and the onset of the ice age that preceded it was associated with a greenhouse cooling.

Having outlined the evidence for the influence of the chemical composition of the atmosphere on climate, I move on to detail the way our own emissions of particular gases may be perturbing the climate.

The book thus far confirms the very real possibility of climatic changes induced by human activity and the second part of the book considers the kinds of climate changes that may occur, and what we might consider doing to limit them. Having read the first part of the book you will not be surprised to discover that predicting what may happen over the next century or so is a great deal more complicated and difficult than it seems, or that our descendants may live in any one of several quite different kinds of climate, not all of them unpleasant.

There is a range of reforms we might undertake that would reduce the quantity of greenhouse gases we release. I suggest that the present uncertainty is such that we should make changes only if they bring incidental benefits, so they are desirable regardless of what may or may not happen to the climate. Perhaps by coincidence, such benefits accrue from most of the reforms that have been proposed.

If we seriously intend to reduce our emissions of carbon dioxide, however, we cannot escape the fact that nuclear power is the only method we have for generating electrical power that both works and that releases no carbon dioxide. In years to come we will have to rely on it much more than we do today.

Unfortunately, for some people the very words 'nuclear power' are frightening. The fear is based partly on the way risks are perceived. Statistically, many studies, based on techniques similar to those used for a century or more by life insurance companies to assess the relative hazards associated with different occupations and personal habits, have shown indisputably that the chance of dying in any particular year from illness caused by radiation released by a nuclear reactor is one in ten million. This is the same as the chance of being killed by lightning, a risk that frightens no one, and far smaller than the risk of being killed in a road accident, which we accept as an unavoidable consequence of the way we live. Yet radiation seems beyond our control. We can shelter in a thunderstorm and so virtually eliminate the risk of being killed by lightning. We can drive carefully and urge others to do the same, so reducing the risk of road accident. Radiation cannot be detected by any of our senses; we may feel there is no way we can influence the design and operation of nuclear installations; and so it may seem that we face a kind of anonymous threat we cannot avoid. This makes it much more menacing and having convinced ourselves of the dangers, assurances from scientists that we are exaggerating them out of all proportion to their seriousness simply add to our unease. Ironically, the safety precautions taken by the nuclear industry give an alarming impression. One commentator has said it is like placing a vehicle on a bathing beach labelled 'Shark Attack Unit'. It would terrify bathers whether sharks were present or not. The fear is not helped by the fact that nuclear generating plants are large, and employ technologies that are poorly understood, because they are seldom explained in everyday language. The fact is, though, that in the thirty or so years since it was first established nuclear power generation has caused no detectable harm to non-human species and, on any assessment, far less injury to human health than its principal rival, coal. If we decide we ought to reduce our emissions of carbon dioxide while sustaining a supply of energy adequate for the continued functioning of the economies of the world - whose failure would most certainly lead rapidly to appalling environmental degradation - nuclear power must be an important ingredient of the mixture of strategies we adopt.

This issue is so important that I have felt it necessary to explain why we should come to accept nuclear power and then to describe briefly what it is, how it works, and what risks it implies.

This is a book about science, and in places difficult science, but I have tried to write simply and to avoid jargon. I live with the uneasy suspicion that ideas which cannot be described lucidly in plain language, so they can be made accessible to anyone, may not be very important ideas. We think in our language, after all it remains our most powerful means of communication, and if an idea cannot be described coherently in language I think it at least possible that the idea has not been thought through in language either, which may mean it has not been thought through at all. Be that as it may, I have assumed no scientific knowledge on the part of the reader, and have tried to explain everything.

You may see the greenhouse effect as a threat. Perhaps it is that, but it is also something else, something I find more interesting. Such as it is, our present understanding of this phenomenon is based on a great web of information gathered from many sources. That information has been assembled in an attempt to explain a particular aspect of the way the world functions. It is an adventure of sorts, an exploration, and in describing it I have tried to capture that feeling. As you read, I hope you may be infected with the excitement of the quest.

CHAPTER 1

The Air We Breathe

The widespread fear, voiced by scientists, environmentalists, and more recently even by politicians, is that human activities may be producing changes in the climate. Put another way, the substances we discharge into the air are altering the Earth's atmosphere in ways that are very likely to prove harmful. Are the fears justified? The question sounds simple enough, but it is a question about the natural world, and the world is a far from simple place.

As we set about unravelling all this complexity, let us begin with a cliché. At some time or other you must have heard our Earth described as a sphere of rock covered, in most places, with the merest smear of water and, outside that, by a slightly thicker layer of gas. That water and that gas, our oceans and atmosphere, contain and sustain everything that lives. To get an idea of the scale, if the Earth were the size of a new tennis ball, that layer, from the bottom of the deep ocean to the very top of the atmosphere, would be less than one millimetre thick. It would be thinner than the pile, the 'fuzz', on the ball.

That is the cliché and like most clichés it is partly true but mainly misleading. It is true in that it demonstrates, in a way we can understand, the thickness of the oceans and atmosphere in relation to the overall size of the planet. It tells us they are rather thin. But that is all it tells us, and taking the comparison any further can make it seriously misleading. A tennis ball can lose its fuzz through wear, until it is quite smooth. A film of dampness on its surface will evaporate in a few minutes if the ball is left in a dry place. The image gives us the impression that the oceans and atmosphere are fragile, not to say ephemeral, but in the real, full-size world, of course they are not. A thin film of water does not behave in the least like an ocean, and when water evaporates from the real ocean it is held by the atmosphere and in due course it returns. The water and air cannot be lost from the Earth, at least not easily - although they will disappear one day, in the distant future.

The idea that all life depends on this film of air and water is misleading in another way. It is true in the sense that we could not live in an airless world or a completely dry one, but we are less helpless than this makes us sound. We, by which I mean all living beings, depend on the air and water, but we also manage it. Indeed, over the billions of years the Earth has been inhabited it is no exaggeration to say that its inhabitants have altered the air and oceans to such an extent that we might well describe them as 'artefacts'; things that have been made, not by people, certainly, but by living beings all the same. We are not 'passengers', riding passively on 'Spaceship Earth', so much as 'crew', maintaining and modifying the 'ship' as our journey proceeds. It is this constant activity that makes life possible. It is also the reason for our present concern. If the 'crew' comprises all living beings, have some of its members - humans - become unruly and irresponsible?

The History of 'Spaceship Earth'

The old-fashioned 'Spaceship Earth' image suggests a world whose life-support systems are inanimate, there since the beginning of time, and fixed. The image is meant to make the world appear fragile but, paradoxically, if it were true it would be quite impossible for mere humans to harm the planet. The oceans contain about 520 cubic km (330 million cubic miles) of water, and the atmosphere nearly 12,600 million cubic km (8,000 million cubic miles) of air. If these huge amounts of air and water were completely inert it is difficult to see how we could affect them in any way at all. Our modest addition of carbon dioxide, raising the amount it contributes to the total atmosphere to its present value of about 350 parts to every million parts of other gases, would have little effect. The air and oceans are far from inert, however, and it is their very dynamism that may render them vulnerable to quite small influences. We need to take a brief look at our history.

The Earth has a history, and so does its atmosphere. The planet formed, along with the rest of the solar system, about 4,600 million years ago, when, inside a swirling cloud of dust and gas, particles collided, stuck to one another, collected others, and so grew by accretion. At first the Earth probably consisted of rocks surrounded by a layer of gas, an atmosphere, composed mainly of hydrogen and helium. As the Sun began to radiate light and heat, that atmosphere was heated and, quite literally, blown away into space. For a time the new planet had no atmosphere.

The Earth grew by sweeping a track through space, gathering to itself any dust and rocks that were close enough to be caught in its gravitational field. This material fell to the surface as an intense bombardment that continued for many years, and as each new arrival crashed, the energy of its motion was converted into heat. You can create this effect yourself, on a tiny scale. If you hammer long and hard at a small piece of metal it (and the hammer) will grow warm. The Earth heated. At the same time the radioactive substances inside its rocks also generated heat - and this heat of radioactive decay is now the main source of the internal heat of the Earth. The hot interior was covered with a cooler, rigid crust of rocks that here and there broke, allowing molten rock - magma - to reach the surface. Volcanoes appeared and eruptions were frequent. In addition to molten rock, volcanoes emit gases, and these gases, belched from the ground and then held by the Earth's gravity, formed the planet's second atmosphere.

At this point you can compare the Earth with its two sister planets, Venus and Mars. The three Earth-like planets are fairly close together within the solar system, are of roughly the same size, and were made at the same time from the same stock of raw materials; and so they should resemble each other closely. All three have atmospheres, and those atmospheres should be much the same.

On Mars and Venus about 95 per cent of the atmosphere consists of carbon dioxide. The rest is mainly nitrogen, with minute traces of oxygen and argon. Presumably, that is what the Earth's atmosphere was like until about four thousand million years ago - which is when life appeared. Today our atmosphere is very different. By volume, it is approximately 78 per cent nitrogen, 21 per cent oxygen, and 1 per cent argon. The 350 parts per million of carbon dioxide comes to 0.035 per cent of the total, and there are very small traces of ozone, hydrogen, helium, neon, krypton, xenon, and methane. About 4 per cent of the atmosphere consists of water vapour but this is not usually counted in the total composition, although it is extremely important, not only in providing our rainfall, but also in connection with the greenhouse effect.

Carbon and its compounds

The change from that old, primeval atmosphere to the new started early, and it started not on dry land, but in water. Perhaps the first living organisms lived in shallow pools where the water was fairly warm and calm, but it was not long before their descendants drifted out into the open sea. They obtained their food from the chemicals in the water, and the element they needed most was carbon. About one-fifth by weight of every living thing on Earth, from the humblest bacterium to the lord of the jungle, not to mention the jungle itself, consists of carbon, (and much of the remainder is water).

Carbon is the basic 'building block' on which life is constructed and there is a very good reason why this should be so: the carbon atom is able to offer four bonds to other atoms. An atom, any atom, consists of a nucleus surrounded by electrons. The nucleus is made up of protons and neutrons. A proton is a particle that carries a positive electrical charge; neutrons carry no electrical charge. The simplest atomic nucleus is that of hydrogen. It consists of just one proton and no neutrons. The simplicity of its nucleus, which is where almost all the mass of the atom is concentrated, makes hydrogen the lightest of the elements, and in the universe as a whole the most common.

Since protons each carry a positive charge, and it is a basic law of physics that like charges repel one another, clearly some other force, stronger than the electrical repulsion, must bind the protons together in all nuclei heavier than hydrogen, the nuclei with more than one proton. There is such a force. It is called the 'strong force' and although powerful, its effects are felt only over the distances encountered inside atomic nuclei. It has no influence beyond the nucleus, where the electrical (or more accurately electromagnetic) force operates. Like charges repel one another but unlike charges attract, and so the nucleus has one or more 'shells' of electrons bound to the nucleus by the electromagnetic force. An electron is a particle much smaller than a proton, but it carries a negative charge exactly equal to the positive charge on the proton.

The number of protons in the nucleus determines the number of electrons there can be in the atom, but adjacent atoms can share one or more electrons. This bonds the atoms together. Two or more atoms bonded together securely form a molecule and if the atoms are of different elements the resulting substance is a compound. Physically and chemically, compounds may be unlike the elements from which they are made. Carbon itself, for example, exists in its pure form either as graphite or carbon black, a soft, almost greasy substance, or more rarely as diamond. Oxygen is a gas our bodies can use to provide energy. Carbon dioxide is a gas we cannot breathe.

Different elements are able to form different numbers of such bonds. There are several ways electrons can be shared, and so there are different types of bonding. Carbon is able to form a maximum of four bonds. This means one carbon atom is able to bind itself to up to four other atoms. The simplest compound to result from such complete bonding involves the attachment of one carbon to four hydrogen atoms. The chemical symbol for carbon is C, that for hydrogen is H, and so the molecule can be described as:

```
      H
      |
  H — C — H
      |
      H
```

or as CH_4. (CH_4 is the symbol for methane.)

Other atoms can take the place of some or all of the hydrogen atoms, other carbon atoms can be added, and so very large, complicated molecules can occur. Take 12 atoms of carbon (C), 22 atoms of hydrogen (H), and 11 atoms of oxygen (O), and you have $C_{12}H_{22}O_{11}$. That is the molecule of sucrose, more commonly known as ordinary table sugar, cane sugar, or beet sugar (they are all the same no matter what the manufacturers may try to tell you).

It is this ability of carbon to form bonds that makes possible the construction of the substances from which living organisms are made, and those first organisms needed carbon. They obtained it from the sea, because carbon dioxide (CO_2 or O—C—O) is slightly soluble in water and, once dissolved, various chemical reactions can separate the carbon and the oxygen and reassemble the atoms in new ways. As the organisms took it into themselves more carbon dioxide dissolved into the sea to replace what they had removed, so indirectly they were removing carbon dioxide from the air.

Some time later a much more effective method emerged. Carbon dioxide dissolves in the sea, but it also dissolves in water droplets in clouds, and so all rain contains a little, making it very weak carbonic acid (H_2CO_3). As the rain flows across the land it penetrates cracks in rocks and reacts with compounds of calcium (Ca), silicon (Si) and oxygen in the rocks, releasing new compounds that flow back to the sea. In particular, calcium is released and the carbon returns as bicarbonate (HCO_3). In the sea, some organisms developed techniques for using these raw materials to make calcium carbonate ($CaCO_3$), a useful material for shells and other rigid structures because it is insoluble near the sea surface (although it becomes soluble at very great depths). When the organisms died their insoluble shells fell to the sea-bed and accumulated. The process removed carbon from the air and deposited it as insoluble sediment on the bottom of the sea. Should you doubt the scale on which this occurred - and still occurs - the evidence is almost everywhere, for some of those sediments, heated, compressed, twisted this way and that by movements of the Earth's crust, have been raised above sea level, where we can see them. They are what we call limestones and chalk, and they are among the commonest of rocks.

When you think of marine organisms that make shells for themselves from calcium carbonate, the animals that spring to mind are likely to be the lobsters, crabs, and other shellfish. These are not the most important groups involved, however, mainly because they are too big. The world can sustain only a limited number of big animals because there is not enough food for them. The vast majority of animals are small, many of them too small for us to see without a magnifying glass or a microscope.

The coccolithophorids are not even animals. They are minute plants, but they use calcium carbonate to make tiny plates that give them some rigidity. The plates are called coccoliths and so numerous and so ancient are the coccolithophorids that they are the largest single source of the calcium carbonate in our limestones and chalks. The white cliffs of Dover are made mainly from coccoliths.

That, then, explains what happened to much of the carbon dioxide in the Earth's original atmosphere. At some point, while all this was going on, another new organism appeared, containing another new molecule with the formidable chemical symbol $C_{55}H_{72}MgN_4O_5$ (Mg is magnesium and N is nitrogen), but the more familiar name of chlorophyll (actually, chlorophyll A, the commoner of the two kinds of chlorophyll found today in green plants). Eventually these tiny organisms insinuated themselves into larger cells, or were consumed by them, and there they live to this day. They are called chloroplasts, and they are what make most plants green, because chlorophyll is green. When light falls on a molecule of chlorophyll some of the energy of the light is absorbed and used to start a long and complex chain of chemical reactions, in the course of which carbon dioxide and water (H_2O) are broken down and reassembled to make carbohydrates, such as sugars and starches. This process, called photosynthesis, led to further removal of carbon dioxide from the air and its accumulation in vast numbers of plants, and animals that feed on plants; a by-product from the chain of reactions is free, gaseous oxygen.

When photosynthesis began there was virtually no free oxygen in the atmosphere. At first the oxygen reacted with exposed substances, oxidizing them, but eventually a time came when there was nothing left to be oxidized and the oxygen began to accumulate in the air. Green plants are the source of all the oxygen in our air and despite the vast forests and our worries about forest clearance, in terms of total mass most of those plants are no larger than a single cell and drift about in the upper waters of the oceans. Clearing the tropical forests is harmful, and perhaps dangerous, but it would be far more dangerous to harm those minute plants, known collectively as phytoplankton (from the Greek *phuton* meaning 'plant' and *plagtos*, 'wandering').

Nitrogen

We have unravelled the mystery of the lost carbon dioxide and the relatively large amount of oxygen, but most of our atmosphere is nitrogen. That, too, was released by living organisms.

Nitrogen is another element that is essential to life. Carbohydrates are all very well but, as children are told when they eat too many sweets, we also need proteins. Proteins are very large molecules made from chains of amino acids. An amino acid consists of carboxyl (COOH) linked to an amino (NH_2). The amino is essential, and it contains nitrogen (N).

The nitrogen came in the first place from volcanoes, as did the carbon dioxide, and it began as an oxide, a compound of nitrogen and oxygen. Nitrogen oxides are soluble in water, so they were washed to the ground by the rain and found their way to the sea. Pure nitrogen gas does not readily form compounds with other elements. Energy has to be applied to it to make it react. Should any nitrogen gas find its way into the air, however, it would not be long before that energy reached it. Over the world as a whole there are thousands of lightning flashes every day. Those giant sparks would supply the energy, the nitrogen would react with oxygen and the resulting nitrogen oxides would dissolve in the rain clouds. So, you see, the air should contain little or no nitrogen.

Dissolved in water, nitrogen oxides end up as nitrates, and nitrates can be used by living organisms. That is the form in which plants absorb their nitrogen. The fertilizers farmers use are spread in a solid form but dissolve as soon as they are wetted. Fertilizers are most commonly based on nitrates, but some are based on ammonium (NH_4) compounds or even ammonia (NH_3). These are also soluble, and in the soil they are converted to nitrates.

Other organisms, though, called denitrifying bacteria, take in the nitrogen oxides, partly to use the oxygen. Some of the nitrogen is released as a by-product as gaseous nitrogen. Denitrifying bacteria are everywhere; they are among the most ancient of organisms, their numbers are vast, and it is they that have put the nitrogen into the air. For billions of years they have been pumping gaseous nitrogen into the air faster than the lightning can convert it to oxides and the rain can remove it. The nitrogen, like the oxygen, has been put into the air by living organisms.

You can see now why our atmosphere has come to be so different from those of our sister planets. Life developed on Earth, but only on Earth, so Mars and Venus still have the atmospheres with which they began, while Earth has an atmosphere transformed utterly by living organisms. Almost all the carbon dioxide has been removed and nitrogen and oxygen have taken its place. Our air is a product of life itself and it is unique, at least in the solar system.

Methane

The air contains very small amounts of methane, another gas about which we will have much to say later. Methane (CH_4) is a very simple compound, but it is not simple to make if you start out with the obvious raw materials of carbon dioxide and water. You have to proceed through a series of chemical reactions and the product of each reaction is very unstable, so there is a good chance the chain will be broken. Methane is produced naturally, however, by bacteria that live and thrive hidden from the air in waterlogged mud and in the airless conditions found inside the digestive systems of animals. Once in the air, methane begins immediately to undergo another series of reactions with oxygen, the end products of which are carbon dioxide and water again. Methane and oxygen cannot exist for very long together because they react and destroy one another. The fact that methane is present in our air is due, yet again, to the activities of living organisms which, until recently, pumped it into the air as fast, but most of the time no faster, than the ordinary chemical reactions removed it. The amount in the air remained fairly constant. It is now increasing.

Earth's Dynamic Atmosphere

I said earlier that if the Earth's atmosphere were at all like the picture of it the 'Spaceship Earth' metaphor suggests, we would have little cause to worry. Now I can explain what I meant by that.

The modern atmosphere of Earth is very different from those of Mars or Venus. I have outlined some of the reasons for this, but there is more. On Mars and Venus there is nothing controlling the chemical composition of the atmosphere apart from the ingredients themselves, which react in entirely predictable ways using energy from the Sun to power the reactions. In other words, the atmospheres are very stable. Should you perturb them, say by adding truly huge amounts of 'alien' substances to them, nothing much would happen. The introduced substances would be incorporated in the overall composition, perhaps they would react with the gases already there, but in time the entire system would settle down either to something very like its present state or to another state that was no less stable. The outcome would not be very dramatic.

Tinker seriously with the Earth's atmosphere, on the other hand, and exceedingly dramatic consequences would follow. Think what would happen, for example, if the amount of oxygen in the air were to increase.

Oxygen is a very reactive element. If you doubt this, think of the ease and speed with which it can combine with the iron in the steel from which your car is made to form a reddish-brown iron oxide - rust. Oxygen reacts readily with carbon, and when the reaction proceeds rapidly, which it does at quite modestly raised temperatures, the reaction liberates energy in the form of heat and light. This is what happens when you burn fuel that is rich in carbon. Whether or not carbon will burn in oxygen depends mainly on the proportion of the air that is oxygen. So if you increase the concentration of oxygen in the air you also increase the likelihood of fires. If the concentration increased from the present 21 per cent to around 25 per cent, all exposed carbon would burn merrily. The forests would burn, the crops in the fields would burn, you and I would burn because we are made largely from carbon, and even the wet seaweed on the beach would burn.

If the concentration of nitrogen were to increase the consequences would be quite different. It is reluctant to react with anything and so it would become much more difficult, and perhaps impossible, for anything to burn at all because the increase in concentration of nitrogen would imply a decrease in the concentration of oxygen.

Any substantial change in the proportions of all the

constituents of our atmosphere would have consequences we would find distinctly uncomfortable. The relationship applies equally in the other direction. If the living beings need the atmosphere to be the way it is, the atmosphere could not remain that way without the management supplied by the living beings.

If all green plants were to die, for example, photosynthesis would cease and no more oxygen would be pumped into the air. The plant tissues would be broken down by the animals, fungi, and bacteria responsible for decomposition and, in a little while, all the carbon they contain would be oxidized, using the oxygen from the air, and return as carbon dioxide to the atmosphere from whence it came all those millions of years ago. If the denitrifying bacteria were to perish, little by little the air would lose most of its nitrogen. Without the living organisms, therefore, the atmosphere could not retain its present composition. The point is that the atmosphere of Earth is maintained in a balance that suits living organisms rather well but that is not particularly stable and it is maintained, somewhat precariously, by the constant activity of those organisms themselves.

The atmosphere is dynamic and because of that dynamism it may be vulnerable. It is possible that humans could make quite small changes that had profound consequences. The concept of the greenhouse effect is based on such a small change.

Physical Structure

So far I have described the chemistry of the atmosphere, but the purely physical aspect is no less important and contributes a dynamism of its own. One result of this is that the air does not form one complete, homogeneous mass, the same from the ground surface to the edge of outer space. It is more like a layer-cake.

When a substance is heated, energy is delivered to the molecules of which it is composed. With more energy, they move faster. If the substance is a solid the molecules may only vibrate, but they do it more vigorously. If it is a fluid, in which molecules move around, they move faster and further and this has the effect of making the fluid expand. When it expands the same number of molecules occupy a larger volume, and this means it becomes less dense. Being less dense, it will be lighter, and being lighter it will rise through the denser, heavier fluid above it.

The air has weight (or more strictly, mass). Obviously it must have mass or it could not be retained by the Earth's gravitational attraction. Being a fluid, however, it can be compressed and it is denser at the bottom, near the ground surface, than it is at higher levels. The weight of the air at a particular point is measured as the 'air pressure'. At sea level it is around 30 lb (weight) on every square inch. As you climb, the pressure, or weight of air above you, decreases. Where the overlying weight is less the air is less compressed and so it is less dense.

The surface of the Earth is warmed when the Sun shines on it and the air in contact with the surface is warmed by that contact. The warm air rises, but as it rises it enters less dense air and this allows it to expand further. This expansion uses up some of the energy it acquired when it was warmed, and so it cools. The rate of cooling is called the lapse rate. In unsaturated air the lapse rate is about 9.8°C for every km the air gains in altitude (about 3.6°F for every 1,000 ft). This cooling of a 'parcel' of air without any exchange of heat with the surrounding air is called adiabatic cooling. It is why the air grows colder as you climb a mountain and why high mountains are capped with snow even in the tropics.

The expanding, cooling air continues to rise until it reaches a level at which its density and temperature match those of the air surrounding it. At this point it can rise no further and so it remains where it is. This sets an upper limit to the lowest layer of the atmosphere, the layer that is dominated by the heating of the air by contact with the ground surface. The layer is called the troposphere (from the Greek *tropos* which mcans 'turning') and the boundary between it and the layer above is called the tropopause. The height of the tropopause varies, but on average it is at about 18 km (11 miles) over the equator and about 6.5 km (4 miles) over the poles. The temperature at the tropopause is about −50°C (−56.5°F).

Above the tropopause lies the next layer, the stratosphere (from the Latin *stratum* meaning 'something laid down'), extending to a height of about 50 km (30 miles). The air temperature does not fall as you climb through the stratosphere until you approach the top, where the air is warmed directly by the Sun. At the upper boundary of the stratosphere, the stratopause, the temperature is about 0°C (32°F). Modern jet aircraft fly in the lower regions of the stratosphere, a few research aircraft can reach the middle, but only spacecraft go beyond the stratopause and into the mesosphere (from the Greek *mesos* meaning 'middle'), where temperatures start falling again until, at the mesopause at about 80 km (50 miles) the air is at about −90°C (−130°F).

Beyond the mesopause lies the thermosphere, where air temperature increases with height. Above about 100 km (62 miles) the gases comprising the air are bombarded by cosmic and solar radiation that strips electrons from its atoms, leaving them carrying an electrical charge. An atom that loses electrons is said to be ionized and so this layer, between about 100 km and 300 km (186 miles) is sometimes called the ionosphere. There are more layers even beyond that. The exosphere begins at a height of some 500 to 750 km (310 to 470 miles) and the magnetosphere extends beyond that until, probably at a height of around 8,000 km (5,000 miles) the Earth's atmosphere merges with that of the Sun. It is difficult to say just where our atmosphere ends, but for most practical purposes it does not extend into these remote outer regions. The air pressure at 50 km (30 miles) is one-thousandth of the pressure at sea level, and by 90 km (56 miles), it is one hundred-thousandth of the pressure at sea level.

The weather we experience is confined to the troposphere, but it is strongly influenced by what happens in the stratosphere. The ozone layer lies in the stratosphere, for example, with a concentration of ozone at around 22 km (14 miles) and, as we shall see later, the ozone layer has a small influence on the climate.

The troposphere is a turbulent place. It is where warmed air rises and cools, colder air descends, and it is also the region where the air contains water vapour. Conditions in the stratosphere are much calmer, although in particular latitudes there are very strong winds, the jet streams, and stratospheric air is very dry.

When warm, dry air moves over the surface of water, water evaporates into it. The amount of water vapour the air can contain depends on the temperature. As the temperature falls the vapour condenses. It is why the cold air in a refrigerator causes water vapour to condense out of the air and form ice. By the time air has moved all the way up to the tropopause it has cooled so much that all its water vapour has condensed. The thin, wispy cirrus clouds you sometimes see, composed of tiny ice crystals, are close to the tropopause and represent the last bit of water frozen out of the air. If you fly through the clouds you hardly notice them, they are so thin and diffuse. You would not be advised to fly through the intense turbulence of a huge cirrocumulus thundercloud, with an anvil at its top. The top of the anvil shape actually marks the tropopause. The cloud can penetrate no further.

The tropopause is a very real boundary. Only the tiniest amounts of water can cross it and what applies to water applies to most other substances that are common at lower altitudes. Dust and smoke particles are trapped in the troposphere and so are the pollutant gases from our homes, farms, cars, and factories. This fact has two consequences. The first is that most substances entering the air from ground level are removed from the air quickly. A molecule of water vapour remains airborne for an average of about nine or ten days. Particles, of dust, smoke, or salt from sea spray, for example, remain in the lower air for about six to fourteen days, and for up to four weeks if they reach the upper troposphere. They are removed by coming into contact with solid surfaces and sticking to them – the way dust and soot can settle on your furniture (and washing) – or by being washed to the ground by rain or snow.

This makes it sound as though pollution of the troposphere is not very important but this is far from true, and neither is it true that just because particles do not remain in the air for long they cannot produce effects at considerable distances. During World War Two, the tank battles in North Africa stirred up so much dust it could be seen, as spectacular sunsets, over the Caribbean. Today about one-third of all the dust in the lower air has been put there by human activities; so large an amount cannot be called insignificant. Were particles not removed so quickly by natural processes the air would be a great deal dirtier than it is; for some years now we have been releasing particulate matter faster than it can be removed, so it is tending to accumulate.

Oxygen and hydroxyl also provide a mechanism for the cleansing of the air. Water is conventionally described as H_2O, but a rather more accurate description is H—O—H and with the application of a little energy one of its bonds can be broken, so it becomes H + O—H. The O—H is hydroxyl, which exists uneasily because the uncompleted bond (—O—H) will attach it to almost anything. The reactivity of oxygen and hydroxyl allow them to form compounds with many pollutants and, the molecule of the compound being larger than that of the original pollutant, this increases the ease with which they can be removed by rain. At the same time, the reaction makes the pollutant molecule itself less reactive; this reduction in the ease with which it will react with, and so affect, other molecules means that if it is poisonous it becomes less so.

Should particles enter the stratosphere, however, they may remain for much longer. There are no solid surfaces to trap them, no rain or snow to wash them away, and just as it is difficult for particles to cross the tropopause upwards and enter the stratosphere, it is no easier for them to cross the other way, downwards. Particles in the upper stratosphere remain there for three to five years. This means that the consequences of pollution will depend on whereabouts in the atmosphere the pollution occurs and in particular on whether the pollutant is in the troposphere or the stratosphere.

One important difference that can influence the climate arises from the fact that stratospheric air is warmed directly by sunshine and tropospheric air is warmed indirectly, by contact with the warmed ground. This being so, if particles in the stratosphere intercept more of the incoming radiation from the Sun the stratosphere will be warmed, but less of the radiation will be able to penetrate all the way to the ground, so the ground and the air above it will be cooler.

Volcanoes

We do know what happens when the stratosphere is seriously polluted, because large quantities of particles may enter it when there is a very large volcanic eruption. The amounts of material ejected by volcanoes can be vast, but most of it falls back to the surface very quickly because the particles are too big and heavy to remain airborne for long.

Eruptions usually leave a thick deposit of dust and ash and our knowledge of volcanic activity in the remote past is based largely on that ash, still forming well defined layers in the rocks. Only a very small proportion of volcanic material can ever reach the stratosphere and once there it is quickly diluted as it spreads over an immense area. Even so, such a thin veil can produce measurable effects. Generally, the effect is to warm the stratosphere and cool the troposphere. In the troposphere itself, most polluting particles have the opposite effect. They warm the air by absorbing heat.

In modern times there have been three truly tremendous eruptions. The biggest was probably at Mount Tambora, in what is now Indonesia, in 1815. It blasted some 55 cu.km (35 cubic miles) of debris into the air and some of the fine dust reached the stratosphere. The following year, 1816, was known as the year without a summer. That was the year in which the poets Percy Bysshe Shelley and Lord Byron, with Mary Godwin and Claire Clairmont, chose to take a holiday in Switzerland. The weather was so appalling that most days they found themselves trapped indoors. You will be glad to hear that they did not waste their time in idle pursuits. Indeed, it turned out to be a rather productive holiday. Mary Godwin began work on what was to become her novel *Frankenstein*, Byron wrote part of his *Childe Harold*, and Shelley and Godwin planned to marry, and did so later.

Another large eruption, immortalized by Hollywood, occurred in the summer of 1883 on the small volcanic island of Krakatau (or Krakatoa) between Djawa (Java) and Sumatera (Sumatra), and therefore to the west of Djawa, despite the title of the movie. There were several small eruptions in May and June of that year, but on August 26 and 27 a series of huge explosions (the biggest was at 10 a.m. on the 27th), heard more than 3,200 kilometres (2,000 miles) away in Australia, hurled into the air nearly 8 cu.km (5 cubic miles) of ash and rock fragments. The cloud blotted out the Sun and the darkness lasted for two and a half days everywhere within an 80-km (50-mile) radius of the eruption. Some of the dust reached the stratosphere where it remained and was carried right around the Earth several times, causing magnificent red sunsets the following year. There was also a general but slight fall in average temperatures which lasted for several years. It cannot be proved, but many scientists believe the cooler weather was caused by the dust injected into the stratosphere from Krakatau.

The March 1963 eruption of Mount Agung, in Bali, was monitored much more closely. Particles from Agung warmed the lower stratosphere over the equator by 6–7°C (11–12.5°F) and this probably led to cooler weather. More recently, cooling of the weather, caused by the eruptions of Mount St Helens in the USA and El Chichón in Mexico, was very minor and brief.

Obviously, the more particles there are the larger the effect will be, but what matters most is what scientists call the column density. This is the total thickness of the cloud of pollutant over a particular place, all the way to the top of the atmosphere. It is an important concept. You can see why on a day when the air is hazy (due to dust). The sky above you is blue, but if you look just above the horizon it is much paler, perhaps almost white, because the light must travel through a much greater thickness of haze if it travels obliquely than if it travels vertically.

On Venus the effect is more dramatic. Venus is permanently blanketed in cloud and nowhere is the surface of the planet visible from space. If you stood on the surface, though, you would see not cloud but a fairly thin haze, and a yellowish sky. The cloud turns out to be haze, but the atmosphere of Venus is so thick that the column density is sufficient to make the haze semi-opaque. There are no stars to brighten the night-time sky on Venus.

As I said earlier, the natural world is a complicated place. It is not too difficult to analyse the air we breathe, or the smoke and steam and gases coming from a factory chimney, but to understand how the atmosphere as a whole came to be the way it is and to understand, even approximately, the processes going on in it and affecting it, are only now becoming possible; and although scientists have learned much their knowledge is still far from complete.

The global climate, which we experience locally as our weather, results from the interaction of two sets of processes. The first, controlled on Earth, produces an atmosphere with a particular composition. The second, to which we living organisms can but react since it is far beyond any hope of our control, occurs on the Sun, our star. The Sun supplies the energy that on the one hand allows life to exist and on the other hand drives the movements of air and water that give us our weather. The next step to understanding the way the climate works must take us deep into the very heart of the solar system.

CHAPTER 2

The Warmth of the Sun

Our Sun, you are sure to have heard, is a very ordinary star. Astronomers describe it as a main-sequence star, of spectral type G2. There are billions of stars just like it in our own galaxy alone, and countless more in other galaxies. (This fact is used to support the idea that many other stars must have planets orbiting them and that some of those planets must support life – an idea we have no way of proving or disproving.) The bald statement is true, and the Sun is a star of a common type, but what does that mean? To answer that I must tell you a story.

The Birth of the Sun

Long, long ago, somewhere along one of the spiralling arms of the island of stars we call the Milky Way, a star was born. It was extremely large and probably it did not last for very long, perhaps for no more than about 12 million years. It died, not in quiet obscurity but in the most dramatic and violent event the universe can produce. It exploded, as a supernova. The intense pressures and high temperatures generated in the events immediately preceding the explosion formed a suite of atoms of the elements heavier than iron, some of them radioactive. The explosion blasted away the entire outer shell of the star, sending it hurtling through space as a hot cloud of gases mixed with those heavy atoms.

About 5,000 million years ago, somewhere else along that arm of the Milky Way, the hot cloud, travelling at great speed, collided with a cool, much more tranquil cloud. The two merged but the collision caused turbulence. The new, mixed cloud began to swirl and twist. Material in it was brought together and held. Little by little it began to collapse in on itself under its own weight, and then it began to rotate about its own centre.

Hydrogen was by far the most common element in the cloud and so about three-quarters of the condensing mass of material consisted of hydrogen. Most of the remainder was helium and the heavier elements accounted for no more than a tiny fraction of the whole.

By the standards we are used to on Earth, the cloud was immense, although by galactic standards it did not amount to anything very special. Once you move away from our small planet measurements of anything at all are likely to produce numbers with more noughts than it is convenient to count. The astronomical unit has been devised to help with distances inside the solar system. One astronomical unit is equal to the average distance between the Earth and the Sun, which is 150 million km (93 million miles). The diameter of the cloud was something like 100,000 astronomical units. The mass of the cloud can be written conveniently as around 2×10^{30} kg. What that means is about two billion billion billion tonnes, where billion means a thousand million (a one followed by nine noughts).

The cloud collapsed on itself under its own weight and, as you can see, that weight was considerable. Almost all of the mass - more than 99 per cent of it - concentrated near the centre but a small amount at the edge of what had become a rotating disc formed small, separate crumbs - the protoplanets, together with the asteroids in a belt where a small planet failed to form, and a residue of gas, ice and rock, some in the form of comets, right out at the furthest edge. The diameter of the central part shrank from 100,000 astronomical units to about 1,400,000 km (870,000 miles), but a core within that central part, with a mere 1.5 per cent of the volume, contained half the mass.

As you can imagine, the pressure inside that core, with the weight of all the material pressing in, was huge. Today it is calculated at around 425 million tonnes per square centimetre (2.7 billion tons per square inch). The material itself consisted mainly of the nuclei of hydrogen atoms - protons, you will recall - from which the electrons had been stripped to remain as free electrons attached to no protons. So great was the pressure that protons were forced together, in what is called a proton-proton reaction. This reaction converts several hydrogen nuclei (protons) into a helium nucleus consisting of two protons and two neutrons (under certain conditions protons can be changed into neutrons and neutrons into protons) and in the process a small amount of matter is lost by being converted directly into energy according to Einstein's famous equation $E = mc^2$, where E is the energy released, m is the mass of the matter that is lost, and c is the speed of light (in this case multiplied by itself). In other words, a small amount of matter is the equivalent of a large amount of energy.

About 4,500 million years ago the conversion of hydrogen to helium and the consequent release of energy ignited a 'furnace' in what by now had become a new star – the Sun. The energy was in the form of gamma and X radiation and neutrinos, strange particles that so rarely interact with anything that they stream out of the Sun and pass right through the Earth as though it were empty space. The gamma and X-rays, however, remained trapped inside the core of the Sun by the free electrons which scattered them, and by the atomic nuclei which absorbed them, and a million years passed before the rays were sufficiently intense for them to work their way to the surface. Then the Sun began to radiate. The planets, meanwhile, had formed and were orbiting. They were warm because of the energy delivered to them by the intense bombardment that had made them and because the radioactive elements within them were decaying.

Radiation from the Sun

When the Sun began to shine, at first it did so only weakly. All the same, it was radiating and its planets were bathed in its radiation, although being so small they were exposed to only a small proportion of it, about one part in 120 million of the total solar output. The energy the Sun delivered warmed them, and from that moment those that had atmospheres either lost them or acquired climates. The weather machine had been brought to life but, as I explained in the last chapter, it was a little while yet before the young Earth enjoyed its benefits.

There are several kinds of radiation. Radioactive substances are composed of unstable atoms which decay, radiating as they do so. This radiation may consist of alpha or beta particles or of photons. An alpha particle is a helium nucleus (two protons and two neutrons) and a beta particle is an electron. The Sun radiates protons and electrons in a stream, called the solar wind, travelling at about 400 to 600 km (250 to 370 miles) per second, whose intensity changes from time to time. These electrically-charged particles are trapped in the uppermost regions of the Earth's atmosphere and do not reach the surface. Intense bursts of them, emitted during solar storms and taking around 50 hours to reach us, sometimes cause the spectacularly beautiful aurora borealis and aurora australis, the northern and southern lights.

The radiation that reaches the lower atmosphere and the Earth's surface consists of photons, the units that comprise electromagnetic radiation. A photon is a little difficult to understand. It has no mass and carries no electrical charge, and so it cannot be said to exist at all in quite the way a solid object exists, although a stream of photons exerts a small pressure. A photon travels, always at a constant speed, although this varies according to the medium through which it is passing. In a vacuum, photons travel at about 300 million metres (about 186,000 miles) per second – the speed of light. So electromagnetic radiation can be pictured as a stream of photons.

It can also be pictured as a stream of waves or pulses, as though the photons were travelling together in bunches, and for our purposes this is the more useful analogy. There is not much you can say about a photon, but quite a lot you can say about waves, and you can measure them.

Imagine a length of rope, secured at one end and with someone at the other end shaking it up and down. Undulations, or waves, will travel along the rope. The distance between the crest of one wave and the crest of the next is the wavelength, and the number of waves passing a fixed point in a particular interval of time (usually one second) is the frequency.

Since electromagnetic radiation always travels at the speed of light, frequency and wavelength are closely linked. The only way more waves can pass in a given time (frequency can increase) is for the waves to be closer together (wavelength decreases). The unit of frequency is the Hertz, abbreviated as Hz, one Hertz being equal to one cycle (the passage of a complete wave, from crest to crest) per second. Radio waves are one kind of electromagnetic radiation. Their frequencies are often grouped together, in wavebands, which can be qualified as long, medium, or short, but the characteristic transmission of a particular radio station is usually given in Hertz, often as megaHertz (MHz, one million Hertz) or kiloHertz (kHz, one thousand Hertz). Wavelength, being a distance, is measured in metres or subdivisions of a metre.

Although photons always travel at the same speed, the amount of energy carried by the radiation they represent can vary widely. This energy is proportional to the frequency or wavelength of the radiation. The greater the energy, the more waves pass in each second and so the shorter the distance from crest to crest.

As the gamma and X-rays travelled to the surface of the Sun they encountered particles, imparted energy to them at their own expense (they lost energy) and so the star began to emit electromagnetic radiation over a very wide waveband, so wide in fact that different parts of it have different names to distinguish them. In theory, though not in practice, the total waveband, the spectrum, is infinitely wide. If we are to describe these sections of the spectrum more precisely we need to introduce a very small unit, the nanometre, (nm), which is equal to one billionth (thousand millionth) of a metre, or 0.000000001 metres.

The gamma waveband includes the shortest-wave energy of all, with wavelengths of around 0.0000001 nm to 0.000001 nm. X-rays are rather longer, with wavelengths of 0.001 to 1 nm (those of longer wavelength are called soft X-rays). Both gamma and X-rays can pass through most materials, but as they do so they impart their energy to the atoms with which they collide. This tends to disrupt the atoms, by stripping electrons from them (they become ionized) and it can damage living tissue. As a result of such collisions the radiation is absorbed. Gamma and X-rays emitted by the Sun and to which we might be exposed are absorbed completely in the upper reaches of the Earth's atmosphere and they never penetrate to the surface.

At the other end of the electromagnetic spectrum wavelengths are much greater. Microwaves occur in a band from 0.1 mm (millimetres not nanometres) to 2 cm and radio waves range from about one metre to 300 metres.

Between these extremes lies the region in which the Sun radiates most intensely. Of all the solar radiation we receive, about 9 per cent is in the ultraviolet band, from about 400 to about 4 nm, about 45 per cent is what we recognize as visible light, from 740 to 390 nm, and about 46 per cent is infra-red radiation, from about 750 nm to one millimetre. The visible light waveband can be subdivided, by wavelength, into the colours of the spectrum, as violet (425 to 390 nm), indigo (445 to 425 nm), blue (500 to 445 nm), green (575 to 500 nm), yellow (585 to 575 nm), orange (620 to 585 nm), and red (740 to 620 nm). Sunlight is most intense at about 500 nm, at the green end of the blue waveband.

Electromagnetism is a form of energy and so all electromagnetic radiation involves the transmission of energy. When that energy encounters the molecules of matter some of it may be absorbed, transferring energy to those molecules, which makes them move or vibrate more vigorously. So far as our nerve endings are concerned, the matter is warmed. Whether or not the energy will be absorbed and the matter heated depends on the wavelength of the radiation in relation to the size of the molecule, and the most effective part of the spectrum is the infra-red. When you sit in the sunshine and it warms you, in fact it is the infra-red part of the spectrum you enjoy. If you sunbathe in order to acquire a tan, it is the ultraviolet part of the spectrum to which your skin reacts. The two are widely separated and so, using appropriate filters for sunlight or special lamps, it is quite possible to be warmed without being tanned, and to be tanned without being warmed.

When a body, any body, is warmed it starts to radiate energy. On a warm, sunny day, hold your hand close to a stone wall that has been exposed to the sunshine for a while and you will feel the warmth. Imagine a body that absorbs all of the energy that falls on it and radiates all that energy away again, so it neither gains energy nor loses it in the long term. There is no such body in the real world, but the theoretical concept is important. Such a theoretical object is called a black body. To some extent the Sun behaves as a black body, so does the Earth, and so do you and I. When the Earth absorbs some of the solar energy falling upon it and is warmed, it radiates, in the infra-red part of the spectrum. The terrestrial radiation flows into space and, provided the amount of energy absorbed and the amount radiated are equal, the Earth will become neither warmer nor cooler. It is a little like accountancy, in that you can apply values to the amounts of energy involved and make up an energy 'balance sheet'.

It is the relationship between the wavelength of electromagnetic radiation and the size of molecules that determines whether or not energy will be absorbed and so provides the key to the early warming of the Earth, its climates and their evolution, and to the nature of the greenhouse effect that may be perturbing those climates. Concern about the greenhouse effect is often presented as a simple environmental issue, but as I mentioned in the last chapter, it is anything but simple; this account of the formation of the Sun, its structure, the source of its energy, and the character of its radiation, is necessary if we are to understand fully how the greenhouse effect works.

As the Sun began to radiate and as the solar wind started to blow, the envelope of light gases that comprised the Earth's first atmosphere was warmed. Its molecules gained energy, moved more rapidly, and because they were so light they began to move beyond the grip of the gravitational field. They 'leaked' into space where the solar wind swept them away to the far reaches of what by now had become the solar system.

The light from the Sun was weak and the warmth slight, for the furnace 'burned' only at the very centre of the core. As it burned the core temperature increased and the matter in the core tried to expand. The outward pressure it exerted was balanced by the inward pressure of matter drawn by gravity and so the Sun became stable. It is called a main-sequence star because this stable phase of its existence will occupy almost all of the time during which it shines. It can remain in this condition until its reserves of fuel begin to run low, and those reserves are still plentiful. The conversion to helium of one per cent of its hydrogen will keep the Sun shining for a billion years. Up to now it has consumed between four and five per cent of its fuel.

Red Giant

This is not to say that the Sun remains always the same. It is stable but not static. As hydrogen is consumed, and replaced by helium, so the furnace expands and the overlying layers grow hotter. This process has continued throughout the history of the star, and will do so until the end. Around four billion years ago the surface temperature of the Sun was some 20 to 30 per cent cooler than it is today.

The temperature at the centre of the Sun now is around 20 million °C. The temperature falls between the centre and the surface, where it is a little over 4,000°C. Then it rises again, to about 500,000°C some 14,500 km (9,000 miles) from the surface and to about 1.5 million °C at a distance of 2,400,000 km (1,500,000 miles). Figures for temperatures and distances have to be approximate because there is no precise way of determining just where the surface of the Sun is. At its temperatures no solid matter of any kind can exist, but neither is most of it made from what you could describe as a gas. The density of the outermost parts of the Sun is much less than that of the Earth's air, but in the innermost core, not much more than 800 km (500 miles) in diameter, where about 99 per cent of the mass is concentrated, the density is about 60 times that of ordinary rocks. What kind of matter can exist with that density and at that temperature? Just particles; protons, neutrons, and electrons mainly, swirling around, releasing energy.

The fact that the Sun is markedly hotter now, and radiates more intensely than it did when life first appeared on Earth is important because, as I shall explain in chapter 4, until now the Earth's climates have remained reasonably constant. True, there have been ice ages and warmer periods, but these amount to quite minor fluctuations in temperature. Clearly, something has prevented the planet from overheating. If we are interfering with the planet's energy balance it would be advisable to predict what the Sun plans to do in years to come.

A point will be reached at which its fuel stocks begin to run low. The fact that so far the Sun has used only a very small percentage of the total amount of hydrogen it contains should not lull us into complacency for all is not as it may seem. Not all of that hydrogen is available to it, or will ever become available, for the furnace exists only inside the core, where the temperature and pressure are high enough to sustain the proton–proton reaction, and much of the hydrogen lies far outside the core.

When enough of the hydrogen in the core has been

converted to helium the proton–proton reaction will slow down for want of free protons. This will cause the core to cool a little and when it cools its outward pressure, due to its attempt to expand, will relax slightly. Gravity will become the dominant force, the overlying matter will press inwards, and the core will contract. That will increase the pressure inside the core, this time to a much higher value than it had before, and the temperature will rise dramatically, to about ten times its present value. When that happens a new reaction will begin. Helium nuclei will be forced to fuse. The generic name for protons and neutrons together is nucleons. The helium nucleus contains four nucleons. When two helium nuclei fuse the product is a nucleus with eight nucleons, and that makes a nucleus of oxygen. An oxygen nucleus fused with another helium nucleus produces 12 nucleons, which makes a nucleus of a carbon atom. The helium will start to be converted into oxygen and carbon.

The higher temperature will be transmitted to the outer layers of the Sun, which will still consist mainly of hydrogen. Being warmer and with no overlying weight to compress them, they will expand. As they do so they will cool again, so an observer in another part of the galaxy will see the Sun change colour from white to red. Its size will increase because of the expansion and the Sun will become a red giant. Its fate after that will be of no concern to the inhabitants of Earth, for the four inner planets (Mercury is the innermost one) will be engulfed. They will become part of the Sun itself, immersed in the solar atmosphere, a diffuse cloud of hydrogen, probably at around 4,000°C. For what it is worth, the Sun will not explode as a supernova because it is far too small. Only very massive stars expire in so impressive a blaze of glory.

The Sun will not become a red giant for a long time. At present it is middle aged and it will be another five billion years before our planet is consumed. Meanwhile, though, it will continue to grow slowly but steadily hotter and brighter. When, at last, it expands and turns red there will be no one on Earth to witness the event, for Earth will have become uninhabitable long ago. Indeed, most life on our planet may be destroyed in about one hundred million years from now because the mechanisms which so far have prevented the planet from overheating will prove unable to cope with the increased solar output. If we are perturbing the global climate it would be much safer to do so by making it cooler than by making it warmer.

Small Changes – Large Effects

I have said the Sun will continue to grow hotter and this is true, but within that general rise in output there will be fluctuations. The Sun is a good deal less constant than it may seem and very small changes can produce quite large effects. A temporary increase or decrease in solar output of one or two per cent would be virtually impossible to measure but it might be enough to plunge the Earth into an ice age, or to end one. The link between solar activity and climate is not proven conclusively, but the circumstantial evidence is so strong that we should at least take account of it.

The Sun rotates as the Earth does but, unlike the Earth, all its parts do not rotate at the same speed. It is made from gas, remember, and is very insubstantial. The high-latitude regions take about 31 of our days to complete one rotation (one solar 'day') and the equator takes 27. The Sun has a magnetic field and one consequence of the way it rotates is that the field changes its polarity (north becomes south and south north) regularly. During the cycle the intensity of the magnetic field increases to a maximum, declines and collapses to a minimum, the polarity reverses, the intensity builds to a second maximum and collapses again, and then the polarity reverses once more. Starting at one polarity change, it takes 22 years for the polarity to return to this state, so there is a 22 year cycle within which there are two magnetic maxima and two minima, 11 years apart. The magnetic maxima produce effects that are visible on the surface of the Sun as dark patches and groups of patches – sunspots.

On no account must you ever try to look for sunspots yourself with the naked eye or, even worse, through binoculars or a telescope. You will not see them that way and you may well be permanently blinded in the attempt. When the Sun is setting and it is red, if you wait until most of the disc has disappeared below the horizon, then you may see sunspots close to the visible edge. Even then it is extremely dangerous to look at them through a magnifying lens.

The spots are regions where the temperature is about 2,000 °C cooler than the surrounding area, which is why they look dark in comparison. They are cool, but they are associated with increased solar activity that represent a small increase in energy output. Do these cyclical changes affect our climate? Studies of records over 300 years have shown that in the western United States droughts are more probable in the two to five years following every alternate sunspot minimum. Droughts, as I shall explain later, are commonly associated with climatic cooling.

The Maunder Minimum

That may not sound entirely convincing but in 1893, after much rummaging through old records, the British astronomer Walter E. Maunder found something he thought remarkable but that remained no more than a scientific curiosity from the time he published it, in 1894 in the journal *Knowledge*, until fairly recently. What Maunder discovered was that between 1645 and 1715 the total number of sunspots reported by observers was less than the number seen in an average year nowadays. This could not be coincidence. Astronomers have been keenly interested in sunspots at least since 1611, when Galileo was one of several observers who reported them. It is not possible that no one was studying the Sun for all those years, nor is it possible that the weather was so bad for all that time that the Sun was always hidden behind cloud. Comets were reported, and the planets were being described. It was odd, but apparently quite genuine, and this period is now called the Maunder minimum. What is more, not only were there few sunspots, there may also have been few aurorae - the northern and southern lights that are associated with increased solar activity, and hence with sunspots - and aurorae are always noticed. There is some doubt about this claim, however, for in 1988 W. Schröder reported in *Nature* that in fact aurorae were reported in central Europe in all but ten years between 1645 and 1705. He summarized recorded sightings that make a total of about 114, 15 of them in 1705 itself, the average number of sightings during any 60-year period being between 60 and 300. In

September, 1989, Victor Loisha, writing from Tomsk, confirmed Schröder's report with his own study of Soviet historical records of 57, or possibly 61, sightings between 1645 and 1715. To this he added Schröder's central European reports, a further 26 from Asia, and concluded that during this 70-year period the average number of sightings each year was similar to that currently observed in Moscow. It does not prove the Maunder minimum did not occur, but it does put in question the use of reports of aurorae to confirm it.

There are other ways the Maunder minimum can be checked, and in the 1970s an astrophysicist, John A. Eddy, set about doing so, mainly because he wished to prove the whole thing was a mistake and the sunspots had simply been overlooked. In fact he confirmed the minimum, and went further.

Between the sixteenth and nineteenth centuries Europeans experienced what is often called the 'Little Ice Age', a prolonged period of unusually cold weather. It is the time when they roasted oxen on the frozen Thames and European glaciers grew longer. The coldest period during the Little Ice Age was around the end of the seventeenth and early eighteenth centuries – the period of the Maunder minimum. Was this coincidence? Eddy went on to compare sunspot records with climate records. In his own words: 'The fit is almost that of a key in a lock. Every decrease in solar activity ... matches a time of glacier advance in Europe; every rise in solar activity ... matches a time of glacier retreat.' (You can read his full account in 'The Case of the Missing Sunspots', *Scientific American*, May, 1977.)

Eddy used the annual growth rings in trees to obtain his information. The rings can be dated accurately and, for those who can read them, they tell what the weather was like during the year they grew. Chemical analysis of the wood reveals the extent of solar activity at the time it formed, because intense solar activity deflects the cosmic rays (radiation from outside the solar system) that in the upper atmosphere cause the formation of radioactive carbon-14. If there is less carbon-14 it means solar activity had increased. In this way he was able to track the relationship back over several thousands of years. He found several other minima and also maxima, each of them lasting from about 50 years to several centuries.

You cannot prove that changes in solar activity actually cause climatic changes on Earth, but if the apparent connection is a coincidence, it is an extraordinary one. The change itself is very small, but small changes can produce large effects. During the period of most intense cold in the Little Ice Age the average temperature was no more than 1°C cooler than it is today.

It may be that we are still emerging from the Maunder minimum and that, although the 11-year solar cycle continues, the general trend is toward a new maximum, with solar activity increasing. It cannot be proved because for rather more than a century now our burning of fossil fuels has been releasing into the air carbon in which much of the carbon-14 has decayed so the carbon-14 record is insufficiently accurate. If we are heading for a new maximum, however, the implication is disturbing. Increasing solar activity is tending to produce a climatic warming, so if our own activities are also producing a warming, the two effects will combine.

The Earth's Orbit

The amount of energy we receive from the Sun depends largely on the Sun itself, but not entirely. It also depends on our distance from the Sun, on the way our planet travels in its orbit.

The time it takes the Earth to complete one full circuit of the Sun is what we call a year. You can picture the Earth proceeding in its stately, very regular fashion and never changing. It is not like that. In the first place, the Earth's orbit does not describe a perfect circle but an ellipse (it is slightly oval). An ellipse has two centres and in this case the Sun occupies one of them. If you divide an elliptical orbit into arcs, each with the same angular width as measured from one of the centres, the sections of the orbit you have marked out will not all have the same length. According to one of the laws of planetary motion worked out by Johannes Kepler (in around 1610), a planet in such an orbit must maintain a constant angular velocity, which means the time it takes to pass through each arc must be the same, so its actual speed must vary. The Earth orbits the Sun, but its speed is not constant. What is more, the orbit itself changes over a 95,000-year cycle. This change affects the amount of solar radiation the Earth receives.

The Earth also rotates on its own axis, and the axis always maintains the same angle to the Sun, but the Earth is not quite upright so the axis of rotation is a little offset. This is partly because the Earth is not quite a sphere – the diameter across the plane of the equator is longer than the diameter from pole to pole. The Moon orbits the Earth and it is the Moon's gravitational attraction acting on the unevenness in the distribution of the Earth's weight that causes the tilt. The tilt gives us our seasons. The hemisphere that is tilted towards the Sun enjoys summer while in the other it is winter. The tilt is not quite steady, however. The Earth wobbles a little (the technical name for it is precession) as it goes along, so the angle of tilt varies between 21.8° and 24.4° with a cycle of about 42,000 years. The bigger the tilt the stronger the contrast between summer and winter. At present the tilt is close to the position it held during the last ice age.

The Earth's rotation is slowing down, so the days are growing longer at a rate of about 35 seconds a century and, because it takes the planet longer to turn on its axis, the number of days in a year is decreasing. Fossils have been found in rocks some hundreds of millions of years old of animals whose growth is controlled by the seasons and it is clear that in their time there were 400 days in a year.

Over another cycle of about 21,000 years, the date of the Earth's closest approach to the Sun changes. At present we are closest on January 3, during the northern-hemisphere winter. This should mean that northern-hemisphere winters are milder than those in the southern-hemisphere and southern summers are warmer than northern summers, but nothing is simple. Air movements over the Earth reverse the situation, helped by the fact that the half year containing the northern summer (March 21 to September 22) is five days longer than that containing the southern summer (September 22 to March 21).

In the 1960s an astronomer called M. Milankovich drew together calculations of changes in the Earth's orbit and records of ice ages and suggested a link between the two that is now accepted by many scientists. According to his calculations, changes in our orbit and precession should produce climatic cycles with periods of 100,000, 40,000, and 20,000 years. This corresponds, more or less, with what is known to have occurred in the past. We know there have been ice ages at intervals over the last 250 million years although they have become more frequent recently. During the last two million years or so there have been about twenty ice ages, with average temperature some 8 to 10°C lower than they are today. If you start by assuming that an ice age is the natural condition for the Earth, at any rate for these last two million years, then periods of warmer conditions (interglacials) occur roughly every 100,000 years and they last for about 10,000 years.

It is worth noting here that at present we are living in an interglacial (called the Flandrian) and that it began about 10,000 years ago. We are due for an ice age any time now. Of course, in any discussion of the history of the Earth 'any time now' really means 'some time in the next few thousand years'. Nor does the Milankovich calculation tell us anything about the extent of the temperature drop associated with the forthcoming glaciation. All the same, an ice age of some sort is where we should be heading. You may not like the idea, but most scientists agree it is what nature intends for us, and 'any time now' could equally well mean 'by next Christmas'.

Perhaps it is time to return to that old image of 'Spaceship Earth'. It may suggest different things to different people but to me it presents a picture of a rather inert Earth, a planet over which we have no control, progressing in a very regular, never-changing fashion in its timeless journey around the Sun. The picture does not include the Sun (how could it, for that would fog the film), but it is there, at least by implication. We can, and I suspect are intended to, infer that our own, private star is benign. It smiles on us across the vastness of empty space, bringing life, warmth and all that is good.

The picture is cosy, sentimental, and wrong. I hope I have done something to discredit it. If the last chapter persuaded you that, far from having no control over the spaceship the living beings of Earth actually constructed its life-support systems and work diligently to maintain them, then in this chapter I hope I have chipped a few edges from the image of a boringly eternal Sun.

The Sun is not eternal. It had a beginning and, at least so

far as its life-supporting role is concerned, it will have an end. It is not constant. There was a time before it bathed our planet in its benign rays and since it first began to shine the intensity of those rays has increased steadily, so now we are bathed in 20 to 30 per cent more of them than our ancestors were four billion years ago. This fact bears strongly on our fears of what we may be doing to our planet through mismanagement for, as I shall explain, this increase in output from a Sun that is beginning to seem a little less than benign required certain measures on Earth to avoid difficulties for the occupants.

Even on shorter time scales, the Sun is not constant, although it was once believed to be so. Its output varies according to regular cycles, and this variation may well affect living conditions for the inhabitants of Earth.

Nor is the stately progression of the spaceship quite so regular as it may seem. The ship is wobbling, slowing down, the shape of its orbit changes, and the date at which it is closest to the Sun changes. To a greater or lesser extent all these factors affect the amount of energy the Earth receives from the Sun.

The picture of Spaceship Earth does include great, swirling clouds, but to most of us they are nothing more than pretty. In fact they are the visible evidence of the meeting of solar energy and the planet, part of the great masses of air whose movements and interactions produce our climates and, on a smaller scale, the weather we experience from day to day.

It is time now to look a little more closely at what happens when the Sun shines and the Earth is warmed.

CHAPTER 3

Where Does the Weather Come From?

Once upon a time, or so the story goes, an examination paper in geography included the question: 'Newfoundland and Britain are in the same latitude. Why, then, is the climate in Newfoundland so much colder than the British climate?' According to the legend, one candidate replied: 'Because Newfoundland is further north'. Like most examination howlers, the report is of dubious authenticity. For one thing the question was not entirely correct. Newfoundland is just a little to the south of Britain.

All the same, if the world were a simple place, two islands occupying approximately similar positions on either side of an ocean ought really to have similar climates, and in this case they very certainly do not. The answer the examiner was seeking, of course, was 'Because Britain is washed by the Gulf Stream'. Probably that would have earned full marks, although it is not correct, either. Strictly speaking, Britain is not washed by the Gulf Stream, although we like to think it is, and in any case the warmish current that does pass British shores is only part of the explanation.

The Effect of the Oceans

The major oceans have gyres, steady currents that flow more or less around their edges. They flow anticlockwise in the southern hemisphere and clockwise in the northern. In the Atlantic, water that is warmed in the Tropic of Cancer flows westward into the Gulf of Mexico, where it acquires the name Gulf Stream. From there it moves northward, along the eastern coast of North America, then turns eastward in the region of Nova Scotia where the continent itself projects eastward into the ocean, southward again in about the latitude of Spain, and so returns to the Tropic along the west coast of Africa. The main Gulf Stream itself comes nowhere near Britain. However, one branch of it does. In the middle of its eastward traverse of the ocean it divides and the branch, called the North Atlantic Drift, flows north-eastwards, part of it around Iceland and part past Britain, along the coast of Norway, round the North Cape, and so into the Arctic. The cold East Greenland Current flows southward down the eastern coast of Greenland, around the southern tip, then north again as the West Greenland Current along the western coast. At the same time, the Labrador Current flows south from the Arctic, along the coast of Labrador. Near Newfoundland the Labrador Current and the Gulf Stream meet, the confrontation between cold and warm water causing many fogs, but Newfoundland itself is influenced mainly by the cold Labrador current. That rather complicated story explains why the climate of Newfoundland is less pleasant than that of Britain.

Clearly, the ocean has a strong influence on climates, and equally clearly it is not enough to know the latitude of a place in order to predict the kind of weather you may expect there. New York is ten degrees south of London, but has much colder winters and hotter summers, and the latitude of Glasgow is the same as that of Moscow, two cities with radically different climates.

There is another reason for mentioning the North Atlantic Drift. One possible effect of a greenhouse warming is that this warm current will be deflected from its present course, so it misses north-west Europe altogether. Should this happen, and it has happened in the past, Britain would become cooler rather than warmer, although it might still be somewhat milder than Newfoundland provided there were no eastward deflection of the flow of the East Greenland current.

The ocean currents are not independent of the air above them. They are driven by the winds. A change in the climate may imply a change in the direction of the prevailing winds and therefore of the currents. This is especially true for the kind of change many scientists are now predicting, in which temperatures remain much as they are at the equator and such warming as occurs is experienced mainly in high latitudes. The winds are one part of the pattern of climate. They affect the oceans, and the oceans in turn also affect the climate. You can see why predicting change is a tricky business.

The climatic influence of the oceans is vast and some aspects of it are poorly understood. I shall refer to the oceans many times as my story unfolds and in chapter 7 I will try to explain a particularly important oceanic phenomenon of which you may have heard: El Niño, and its opposite, La Niña.

I have mentioned 'climate' and 'weather' and it may seem that the two words are synonymous. Indeed, many people treat them as though they were. They are distinct, however, and the distinction is rather important. The weather is what we experience from day to day. It is the subject dealt with in weather forecasts which tell us whether to expect rain or sunshine, warm weather or frost, breezes or gales, over the next few hours or days. The weather is very variable. The climate is a kind of summary of the type of weather a place, or large region of the world, actually experiences over many years and it is a much more reliable indicator of the general type of conditions, and of vegetation, you may expect to find. You can say, for example, that southern France enjoys a Mediterranean climate. This sounds obvious on geographic grounds, but parts of southern Africa also have a Mediterranean climate, which may be less obvious. It is a climate type. It would sound very odd if a weather forecaster said 'tomorrow the weather will be Mediterranean'. Since any climate can bring a variety of weather conditions, the statement could mean almost anything short of the sea freezing.

In a still more general way, the climate is what happens when the warmth of the Sun interacts with the atmosphere and the surface waters of Earth. This makes it into a process rather than a single phenomenon.

The Sun warms the surface of the Earth, the air in contact with the surface is warmed, and so it rises. The Earth is warmed most strongly around the equator, because this is the region that is always presented most directly to the Sun, and at the equator a large part of the Earth's surface is covered by ocean. Water is plentiful and some of it evaporates into the warm, equatorial air and is carried aloft with it. This humid air rises to the tropopause and then drifts away from the equator, northwards in the northern hemisphere and southwards in the southern. Almost no tropospheric air and only a very limited amount of stratospheric air can cross the equator – an important point when considering the transport of pollutants.

As it rises, the equatorial air cools by expanding (adiabatic cooling). The amount of water vapour which air can hold depends on the temperature of the air, so as it cools it becomes saturated. Water vapour condenses to form clouds and rain falls, giving the equatorial regions their very high precipitation. The air is now much drier and, being close to the tropopause, cooler, and it descends again because the air below it is being warmed, and made less dense, by contact with the warm ground. This movement, of moist air upwards and away from the equator and cooled, drier air downwards to the north and south of the tropics, produces a belt of low-latitude deserts around the world. This mainly vertical air circulation forms a kind of cell, called the Hadley cell, after G. Hadley, the scientist who first described it in 1735.

Hadley described one feature of the global behaviour of air and Gaspard Gustave de Coriolis (1792–1843), a French engineer and mathematician, described another, called the Coriolis force, caused by the rotation of the Earth. The air would move precisely north and south were it not for the Earth's rotation. The Coriolis force acts at right angles to the direction of motion of the air, deflecting it to the right in the northern hemisphere and to the left in the southern. Since it is caused by the Earth's rotation about its axis, the force is strongest at the equator, where the surface moves most rapidly, and weakest at the poles. So the air is twisted as it moves and it is this twisting in opposite directions to either side of the equator that effectively divides the equatorial air into two bodies, to the north and to the south, which are prevented from mixing.

When air expands it becomes less dense and the pressure it exerts falls, so the equator is a region of low pressure. As it descends again its pressure is higher, so to either side of the equator there are regions of mainly high pressure. Technically, areas of low pressure are called cyclones and the weather associated with them cyclonic, and the corresponding terms for areas of high pressure are anticyclones and anticyclonic.

It is the fact that the Earth's axis of rotation is slightly offset that produces seasonal changes in the weather (not climate!) outside the equatorial belt. The thermal equator, the belt that receives the most solar radiation, is usually a little to the north or south of the geographic equator. When the tilt in the rotational axis increases the exposure to the Sun of one or other hemisphere the thermal equator moves into that hemisphere, which then enjoys its summer. In effect, the whole climatic system moves a little to the north or south. The tropical rains move into a higher latitude, giving most subtropical regions their rainy season in the summer, and the belt of dry, descending air moves into higher latitudes still, bringing warm, dry conditions to the middle latitudes. In the highest latitudes the day length increases greatly – the Arctic Circle is defined as that region within which there is at least one day in the year during which the Sun does not set fully and one day in which it does not rise fully. The increased radiation due to the greater day length is the main cause of seasonal changes in these regions.

Air tends to flow from areas of high pressure to areas of low pressure but it cannot do so directly because of the Coriolis force. Instead the air spirals, and in the northern hemisphere it moves anticlockwise around centres of low pressure and clockwise around centres of high pressure. In the southern hemisphere these directions are reversed. When lines are drawn on a map to link places where the air pressure is the same they often form approximately circular patterns around centres of low and high pressure and the winds flow almost parallel to the lines. Such lines resemble contours and are called isobars (from the Greek *isobares* meaning 'of equal weight').

The flow of air from the high-pressure areas between latitudes 30° and 40° produces the steady winds - called trade winds by the mariners who found them valuable in the days of sailing ships - that blow around latitude 15° north and south, from the north-east in the northern hemisphere and from the south-east in the southern. The trade winds occur only in these particular regions but they have a significant influence on the global climate.

In polar regions the air is very cold and therefore dense, so there are areas of high pressure in the Arctic and Antarctic. Between the areas of more or less stable weather conditions in the tropics and subtropics and the polar regions in both hemispheres there is a belt where the two types of climate meet. These belts enjoy temperate conditions but because at any time they are liable to fall under the influence of one or other major regime their weather is changeable. Britain lies in the temperate regions and, as we know from our own weather forecasts, the weather is very difficult to predict even for as little as a few hours ahead.

A picture begins to emerge of vast masses of air, constantly on the move, rising and descending, colliding with one another, moving past, merging, and warming and cooling as they go. The characters of these air masses are determined partly, but only partly, by the latitude from which they come, that part of their origins which we can use to label them Arctic or tropical. They are also affected by the amount of moisture they carry and that, in turn, depends on whether they have passed over the ocean to acquire moisture or over continents where they lose it. So air masses can also be continental or maritime.

The Albedo Concept

The characteristics of continental and maritime air are radically different from one another and the difference amounts to more than the amount of water each carries. We have to introduce another very important concept, that of albedo, or reflectivity.

In summer, most people prefer to wear light-coloured clothes because they believe light colours will reflect the heat and so keep them cooler. They are applying the albedo concept, although in the case of clothing it is only partly effective. Summer clothes keep us cool mainly because they are made from thinner materials and we wear fewer layers of them, but it is true that they reflect more radiation than do the dark-coloured clothes we wear in winter because they absorb heat.

What is true of our clothes is equally true, and on a much larger scale, of the planet itself and the Earth's albedo varies greatly from one type of surface to another. As the radiation we receive from the Sun passes through the atmosphere to the surface, about 30 per cent is reflected and the remainder absorbed, but the effect is far from even. Fresh snow, for example, reflects 80 to 90 per cent of the radiation falling upon it, and is said to have an albedo of 0.8 to 0.9. Sand has an albedo of 0.30 to 0.35 (it reflects 30 to 35 per cent of the radiation falling on it), grass or green farm crops 0.18 to 0.25, and forest from about 0.07 to about 0.18 depending on whether it is broadleaved, evergreen, or coniferous. Water surfaces have an albedo of 0.06 to 0.10, although this increases greatly when the Sun is low in the sky and you can see it shining on the water - and therefore being reflected. What is not reflected is absorbed and so the oceans absorb up to 90 per cent of the radiation they receive, fresh snow absorbs about 10 per cent, and other surfaces absorb amounts between these two extremes. Very small solid particles, called aerosols also alter the planetary albedo. I shall return to their influence later.

The low albedo of the oceans makes them a very effective 'heat sink' and this quality is increased greatly by the transparency of water and the great volume and turbulence of the deep oceans. Because water is transparent to solar radiation in the predominant wavelengths (it is much less transparent to shorter wavelengths, including those in the ultraviolet part of the spectrum) about 20 per cent of the light and heat penetrate to a depth of up to 9 metres (30 feet). The turbulent mixing of water caused by waves and currents carries warm surface water still deeper. In the North Sea in summer the warmth of the Sun can be detected down to about 40 metres (130 feet). The mixing also cools the surface waters so they can be warmed again.

Specific Heats

Water also has a high specific heat. The specific heat of a substance is the amount of energy that must be supplied to it in order to raise its temperature by 1 °C. You need not worry about the units employed, but only about the numbers which allow us to make comparisons. The specific heat of water is 4.18 joules per gram and it means that water absorbs a great deal of heat with only a small change in its temperature. It is why, even in the middle of a really hot summer, the seas around northern European coasts remain cold and why in some years the sea can be warmer in the middle of winter than it is in summer. The hardy souls who take a rapid dip on Christmas Day may be not quite so brave as they look because the sea is often warmer than the air through which they must run to reach it!

The result is that the oceans absorb heat but warm only very slowly. In winter they also cool very slowly because they lose their warmth mainly by mixing with cooler, deeper waters - the process that made their warming such a slow process. Obviously, this effect is much less pronounced in shallow waters, and in waters that are almost entirely landlocked and undergo little mixing with larger bodies of water.

Land surfaces, in contrast, are opaque and are not subjected to turbulent mixing. They reflect more radiation than the oceans, but what they absorb warms only the surface layer. Being relatively shallow, this layer warms rapidly. If you measure the temperature at the soil surface on a hot summer's day, then make a hole and measure the temperature again a foot or so below the surface you will see how rapidly the temperature falls with increasing depth. The rate of fall will depend on whether the soil is wet or dry and on the type of soil. Light, sandy soil, for example, contains many air spaces, air is a poor conductor of heat, and so the soil heats rapidly because warmth cannot be dispersed by conduction. The effect is quite dramatic. One hot day in Japan the temperature drop was measured. A sandy soil was found to be at 40°C (104°F) at the surface, and at 7°C (45°F) at a depth of 15 cm (6 inches).

The specific heat of soils is also much lower than that of water. In the case of sand it takes only 0.84 joules of energy to raise the temperature of one gram by one degree Celsius. In other words, the temperature of a given amount of sand can be raised using only one-fifth the amount of energy that would be needed to produce the same warming in an equivalent amount of water. Land surfaces warm quickly and, for the same reason, they cool quickly. The oceans warm slowly and cool slowly.

The difference in albedo profoundly affects the air passing across different surfaces. When air passes over snow-covered regions it will be chilled because snow reflects almost all the radiation falling on it and so remains cold. The continental land masses warm quickly in summer but cool rapidly in winter, so continental air masses are usually very hot in summer and bitterly cold in winter. The temperature at the surface of the oceans changes only very slowly so there are no extremes and little difference between summer and winter. Maritime air masses tend to be cool but not cold and not much warmer in summer than they are in winter.

Clouds and Temperature

Cloud also affects the temperature at the surface, but the relationship is more complex than it may seem. We all know that on a sunny day we feel warmer than we do on a cloudy day because the cloud acts like a parasol, shading us from the Sun, but the extent to which it does so depends on the thickness and type of cloud. As you will know if you have travelled by air above the cloud tops, some clouds have a very high albedo. Cumulonimbus, for example, associated with thunderstorms and most common in the tropics, has an albedo of about 0.9, making it as reflective as freshly fallen snow. High level cirrus clouds, on the other hand, because they are so much thinner, have an albedo of about 0.5.

So clouds have a cooling effect but, paradoxically, they also have a warming effect because they absorb radiation in the longer wavelengths. This is mainly radiation from the Earth's surface and it explains why, in winter, nights when skies are clear are much colder than cloudy nights. This double effect, of reducing the amount of incoming radiation by reflecting it back into space and absorbing outgoing radiation and so warming the air, will play a critical but still poorly understood part in any climatic change associated with the greenhouse effect.

The formation of clouds also has a warming effect. When water changes its state - its phase - from solid to liquid or liquid to gas, heat is absorbed as the molecules acquire the energy to move more freely. When water changes phase in the other direction molecules lose energy and heat is emitted. This is the latent heat of water. It varies with temperature, but at 0°C it takes 335 joules of energy to convert one gram of ice to water and 2,500 joules to convert one gram of water to water vapour. When water vapour condenses or water freezes the same amount of energy is released, as heat. This is why it feels a little warmer when cold, dry weather gives way to snow, and why it feels colder when the ice thaws. Ice can also vaporize directly. This is called sublimation and it requires 2,835 joules per gram of ice, the sum of the latent heats of melting and vaporization.

In an intense hurricane the condensation of water vapour in huge, towering clouds creates very warm air which rises rapidly, twisting around a centre where the pressure is very low. At the top of the storm the air cools and is drawn down into the eye to replace the rising air around the sides and as it descends its pressure rises and so it warms adiabatically, making the storm self-sustaining. To a large extent it is the latent heat of condensation that 'fuels' the hurricane. One predicted consequence of a general greenhouse warming is that warmer seas in low latitudes will cause more water to evaporate and then to condense, and that this will produce tropical hurricanes more frequently and the hurricanes themselves will be more intense.

As air masses move, therefore, their temperatures are strongly influenced by the amount of water they carry, that influence being complicated greatly by the fact that the water is constantly condensing and evaporating again.

When as much water has evaporated as the air can carry at its particular temperature the air is said to be saturated. Put another way, its relative humidity is 100 per cent. This is the ratio of the actual water vapour in the air to the maximum possible expressed as a percentage and it is the figure most commonly given in weather reports. When the relative humidity exceeds 100 per cent you would think, by definition, that water would start to condense and clouds would form. This is not what happens, however, and air is quite often supersaturated. Water vapour condenses only with great difficulty unless it can do so on to a surface and in air with a relative humidity of 100 per cent it can form clouds only if there are solid particles of a particular size, called cloud condensation nuclei available. Without condensation nuclei the air must be supersaturated before cloud can form.

There is rarely any problem over land, where the air always carries some dust particles which serve the purpose quite well, but how can clouds form over the remote regions of the oceans? Once a water droplet has condensed on to a particle that particle is sealed inside the water and there it must remain unless the droplet should evaporate again. It stands a much greater chance of being carried with its droplet up and down inside the cloud, colliding with other droplets to form bigger droplets, until the drops are so heavy they fall, carrying the particle back to the ground – although there are still large gaps in our knowledge of the way minute droplets form raindrops. For this reason dust particles are removed from the air in their passage over the sea and far from land they become exceedingly rare.

Over the oceans some condensation nuclei are supplied by sea spray. This throws drops of salt water into the air, where the water evaporates leaving behind tiny crystals of salt. Ocean clouds are produced mainly, however, as a result of the activities of living organisms. Many species of phytoplankton, but especially the coccolithophorids I mentioned earlier, release a chemical compound called dimethyl sulphide, or DMS ($(CH_3)_2S$), where S is the chemical symbol for sulphur. Some DMS dissolves in the sea but some escapes into the air where it undergoes a series of oxidizing reactions that end with sulphate (SO_4). Sulphate exists as tiny particles (aerosols), less than one-thousandth of a millimetre in diameter and that is the right size for the condensation of droplets. Once in contact with water, the sulphate dissolves, eventually to form sulphuric acid (H_2SO_4), one cause of 'acid rain'. Clouds moving in from the sea, where in recent years the phytoplankton has been increasing markedly, are a much more important cause of this form of acid rain than emissions from coal-fired power stations. However, this does not mean the acid rain is not linked to pollution – for it is possible that pollution is helping to cause the increase in phytoplankton.

The relative humidity of air will increase if the air is cooled by being forced to rise. This is why mountainous regions exposed to generally moist air have a high rainfall on the windward side, and a 'rain shadow' forms in the lee of them. It also happens at the boundary between two masses of air when warm air either overrides or is undercut by colder, denser air, and a front forms.

The Movement of Air Masses

Because of the Earth's rotation, air masses tend to move from west to east in the northern hemisphere, and from east to west in the southern, although with many northward and southward deviations along the way. As they move their characteristics change as they pass over land and sea, over relatively warm and relatively cool surfaces and as water evaporates and condenses within them. In winter, for example, cold, dry, continental air moves from Canada across the Atlantic. As it moves it is warmed by contact with the ocean, including the Gulf Stream, water evaporates into it, the warm, moist air rises, clouds form and variations in air pressure cause wind speeds to increase, and by the time the air reaches Europe it has been transformed into maritime air. Western Europe is exposed mainly, though not all of the time, to maritime air.

It is exposure to the ocean that gives an air mass its maritime characteristics so that no matter what its original type may have been, by the time it has crossed a major ocean it will have been altered. Similarly, any air mass that crosses a large continent will be continental air by the time it leaves. Over the oceans and continental interiors, therefore, the air masses almost always have the same characteristics because it is the oceans and continents themselves that produce them.

The characteristics of a mass of air change as it moves and they also change with height. Obviously, with increasing height the air pressure decreases and the air becomes drier, but because of friction between the bottom of the air mass and the ground the air is usually moving at different speeds at different heights so the air affected by the temperature and amount of water at a particular place on the surface does not form a neat, vertical column over that place. As air rises or descends there is also a horizontal component to its movement. Surface friction also affects wind speed. Usually, wind speed increases with height all the way up to the tropopause, although under some weather conditions fast-moving air may descend to the surface.

Close to the tropopause, at around 30° latitude and moving from west to east in both hemispheres, are the jet streams. These are narrow bands of air, moving at 160 km per hour, and in winter sometimes moving at as much as 500 km per hour, formed where the tropical climatic regime meets the polar regime, and associated with a sharp difference in temperature from one side to the other, the warmer side being the one nearer the equator. Their strength and precise location vary constantly and they describe a rather wavy course.

Human Influences on the Global Climate

Despite this generally latitudinal movement, the overall effect of the movements of air and water is to transfer heat from the tropics to the poles. It is the difference in the amount of warming at the equator and poles, the thermal gradient, that produces the particular characteristics of our climates. Most forecasts of a greenhouse effect predict there will be warming in high latitudes while low latitudes remain more or less at their present temperatures, so the thermal gradient will be weakened. Should this happen it seems reasonable to suppose that the weather systems themselves will weaken. It does not follow that this will lead to generally blander weather, with an absence of extremes and increased stability. Indeed, something approaching the opposite may happen. Such a weakening might produce sustained periods of weather of a particular type each of which gives way to the next in a brief spasm of very violent change. It does suggest, though, that wind speeds will decrease overall, which may be bad news for those who advocate wind energy as a significant source of electricity generation. As I explained earlier, such a general weakening of winds may be accompanied by an increase in the frequency and intensity of tropical hurricanes. Hurricanes are not caused by the thermal gradient but are very local.

The system is vast, extremely complicated, and it is driven by energies that so far exceed anything humans could conceivably generate that you might be forgiven for thinking there is no way our activities could cause more than the slightest local perturbation. The global climate appears very stable. We do cause local changes, of course. The climate in large cities is markedly different from that in the countryside surrounding them. The buildings deflect the wind, causing eddies but generally reducing wind speed, so cities are less windy and that makes them feel warmer. They are also made warmer, and drier, because the heat they release, from chimneys, the open doors and windows of heated buildings, vehicle engines, and countless other sources, warms the air and increases the amount of water vapour it can hold without becoming saturated, and because the drainage system means water is not allowed to stand on the surface to be removed by evaporation.

We can alter the climate in the countryside, too. The clearance of forests leaves the land more exposed to wind, so wind strength increases. This has a drying effect as water evaporates and is carried away at once, before the air can be saturated, and the cloud forms somewhere else. Planting forests has the opposite effect.

Such changes are made locally and their effects are local. Is it possible, though, that we might cause perturbations on a much larger scale; that we might actually alter the global climate itself? Our main concern is the greenhouse effect, which is one kind of change, but there are, or could be, others.

Perhaps we could alter the planet's albedo. The clearance of forests from a very large area might achieve this, especially if the tropical rain forests were to disappear. Tropical rain forest has a very low albedo of 0.07 to 0.15 (it absorbs from 85 per cent to 93 per cent of the radiation it receives). Grass and cereal crops have a somewhat higher albedo, of 0.18 to 0.25 and bare soil has an even higher albedo, of about 0.3. Such a change in surface vegetation would mean the affected areas reflected more radiation and absorbed less. This might have a cooling effect in the tropics. Unfortunately, it is not the remedy for greenhouse warming that it may seem, for two reasons. The first is that the forests trap rainfall, reducing evaporation, and release water mainly by transpiration, after it has passed through trees. Remove the trees and the rate of evaporation would increase rapidly just as the warming of the air over land decreased because of the increased albedo, so conditions would become generally drier. This would reduce cloudiness and its part-warming, part-cooling effect. The second reason is that the problem of global warming - if problem it be - arises partly from the reduction in the thermal gradient between low and high latitudes that results from high-latitude warming. Cool the lower latitudes and the gradient is reduced still further. For this reason alone - and there are others! - it would be wiser to preserve the rain forests.

There is another way in which we might alter the planetary albedo. Very small, solid particles - aerosols - in the air can have this effect, but its global consequences are not easy to predict.

I should explain the word aerosol. It is the name given to

a dispersion of a solid or liquid in a gas, in the form of particles that are so small they are hardly affected by gravity and can remain suspended for a long time. An aerosol can produces aerosols, but the aerosols are the particles, not the can itself. The can contains a propellant, a substance that is a gas at ordinary temperatures and pressures but that is sealed inside the can under sufficient pressure to change it to a liquid. The useful substance in the can - paint, hair spray, deodorant and so forth - is mixed with the propellant. When you press the button a small hole is opened through which the pressurized propellant can escape, vaporizing as it does so and carrying the useful substance with it, now dispersed as tiny particles in a gas.

The aerosols in the air consist of soot, ash, salt from sea spray, dust from volcanic eruptions or raised by the wind, sulphate particles, ice crystals, and many other things. They do not remain long in the troposphere because although they are too small to fall rapidly under their own weight, they are soon washed out by rain or collide with surfaces to which they adhere. Aerosols that enter the stratosphere, on the other hand, may remain there for years because there is no mechanism for removing them more quickly.

Aerosols intercept radiation, but the effect they have depends on their size in relation to the wavelength of the radiation. If they are comparatively large, like soot particles, they absorb radiation. This warms them and so the air around them is warmed. If they are very small, they scatter radiation in one of two ways - called Rayleigh scattering and Mie scattering, after Lord Rayleigh and G. Mie respectively. If the diameter of the particles is less than about one-tenth of the wavelength of the radiation, the photons comprising the radiation simply bounce off them. Oxygen molecules are much smaller than this, with diameters about one-twentieth the wavelength of red visible light and oxygen in the atmosphere reflects radiation in this way, but mainly at wavelengths greater than green in the visible spectrum, so it allows blue light, and radiation with wavelengths shorter than the blue, to pass. This is why the sky is blue. Some dust and smoke particles can scatter blue light but allow the red to pass, and they produce red sunsets. Mie aerosols absorb the photons that strike them but then re-emit them at a different wavelength and mainly in a forward direction.

Obviously, any attempt to calculate the effect of aerosols must take account of both types of scattering. It turns out that the effect depends largely on the albedo of the surface below. If the albedo is low, as it is over oceans and forests, for example, the forward – Mie – scattering has little effect because much of the incident radiation is absorbed by the surface anyway and the aerosols are not adding extra radiation. What matters is the amount of radiation they reflect. This never reaches the surface and so the overall effect of the aerosols is to produce a cooling. Where the surface albedo is high, the opposite effect occurs. Much of the incident radiation is reflected by the surface, so the amount reflected away from the surface by the aerosols is insignificant – it would be reflected anyway. However, the radiation reflected by the surface encounters the aerosols which reflect some of it back to the surface again, so it bounces back and forth. This traps heat in the atmosphere and produces a net warming effect.

Translate this to what happens in the real world and you find that the regions of low albedo tend to be found in low latitudes, and especially in equatorial regions. The effect of aerosols will be to produce a cooling in those regions. Regions of high albedo tend to occur in higher latitudes, especially but not exclusively in the Arctic and Antarctic where there are large areas of ice and snow. There is ice and snow at much lower latitudes in winter, and light coloured soils and plants – the bright greens of early spring, for example – also have a fairly high albedo. The result is that these areas will be slightly warmed. If the greenhouse warming is concentrated in high latitudes, so weakening the thermal gradient, then atmospheric aerosols will contribute to it because they will cause a cooling in low latitudes and a warming in high latitudes.

This is true only for aerosols in the troposphere. Aerosols that enter the stratosphere are a long way from the Earth's surface and their overall effect, from both types of scattering, is to cool the troposphere below them, either by reflecting incoming radiation back into space or by trapping it within the stratosphere and so preventing it from penetrating further. Indeed, some people have suggested injecting large numbers of aerosols into the stratosphere to produce just this effect. Unfortunately, we still know too little about the greenhouse effect and the precise numerical relationship between quantities of aerosols and temperature to make this a wise move.

As I have said, aerosols occur in the atmosphere naturally. They are also there as a direct result of human activities and today our contribution amounts to about one-third of all the dust in the air, not to mention the sulphate, ash, smoke, and other particles we release.

Dust is the biggest source of aerosols and it comes mainly from farms. When the wind blows across bare, dry soil – ploughed, or perhaps sown and waiting for crops to germinate – some soil particles are carried away. I have seen this happening even on heavy, clay soils. On light soils blows can be dramatic, reducing visibility greatly. Where the overgrazing of poor, semi-arid lands leaves soil exposed, it will blow readily. If aerosols have an effect, then already we are adding to it significantly.

In this chapter I have tried to give a broad, general account of the way in which the radiation we receive from the Sun affects the oceans and atmosphere to produce the world's climate types. The system is extremely complicated and, in the technical sense of the word, it is also complex. A complicated system is one we could understand if we had sufficient information about it, and we could use that understanding to make accurate predictions of the way it would behave under given circumstances. A complex system is one we might understand in principle but about whose behaviour we may never be able to make precise predictions because it is extremely sensitive to quite small perturbations. It is what mathematicians call chaotic, and it was from the study of weather patterns that the now fashionable theories of chaos emerged, with the so called butterfly effect, in which the flapping of the wings of a butterfly causes tiny air movements that are transmitted and amplified until they cause a major change in the weather later, thousands of miles away.

It is far from easy, then, to make precise statements about detailed aspects of the way the atmosphere works. Just as my outline has been very general, so predictions of the greenhouse effect are very general. There is much we need to learn before the predictions can become more reliable, indeed before we can know certainly whether a greenhouse warming effect will occur at all. Even then the precise consequences may be subject to chaos, and so unpredictable. What we *can* do, still in a general way of course, is to examine the way climates have changed in the past and the possible reasons for those changes. That is what I shall try to do in the next chapter.

CHAPTER 4

The History of the Climate

The Moon is not very far from Earth and receives the same amount of solar radiation. It has no atmosphere and therefore no climate, but it does have a temperature, which you have to measure on the surface rather than above it. A lunar day lasts for 708 hours, or 29.5 of our Earth days, and so there is ample time for the surface to warm by day and cool by night. The daytime temperature at the surface is around 110°C (230°F). By the end of the lunar night the temperature has fallen to about −170°C (−274°F), so the average temperature (half the sum of the two temperatures) is about −30°C (−22°F). You can calculate this, rather than measure it directly, from the amount of radiation the Moon receives from the Sun, the solar constant, which is about 1,360 watts per square metre, minus the amount of radiation that is reflected due to the Moon's albedo.

The same calculation performed for Earth produces an average surface temperature of about −23°C (−9.4°F). This is not very different from the lunar temperature, the Earth being somewhat the warmer due to differences in albedo between the two bodies. Obviously this is not the temperature of the world we know and in fact such a temperature could not occur because at −23°C the oceans would be covered by ice, which would increase the albedo greatly, so the Earth would be still colder. The actual average temperature at the Earth's surface is about +15°C (59°F), a full 38°C (68.4°F) warmer than the simple calculation suggests, and the Earth is not subject to the wild daily fluctuations in temperature that must alternately bake and freeze the poor old Man in the Moon.

Since the Earth and the Moon are at the same distance from the Sun, the only feasible explanation for the difference in their temperatures must be the presence of oceans and an atmosphere on Earth and their absence from the Moon. Because they warm and cool slowly, the oceans damp down the fluctuations but that does not explain the difference in average temperature. The oceans may cool slowly but they must cool eventually and the temperature should fall to something close to the expected value. That leaves the atmosphere as the 'blanket' that keeps us warmer than the Moon.

Venus and Mars

Mars and Venus also have atmospheres, but no oceans. Their solar constants are different from ours because they are not at the same distance from the Sun, but these constants can be calculated and the temperatures predicted. For Mars they predict an average temperature of around −25°C (−13°F) and for Venus about −48°C (−54.4°F) at the top of the clouds, which is the only surface visible from elsewhere in space. Venus appears cooler than Mars because despite being closer to the Sun its total cloud cover gives it a very high albedo. Instruments have now measured the surface temperatures of both planets and we know that the average temperature on Mars is about −53°C (−63.4°F) and on Venus it is 477°C (890.6°F) which is about 100°C hotter than the melting point of lead. Conditions on Venus are as bad as this temperature makes them seem, but those on Mars are not quite so bad because the average conceals wide differences. At midday at the martian equator the temperature can sometimes be as high as 10°C (50°F), though it drops rapidly at night and in higher latitudes it can fall to around −140°C (−220°F).

Temperatures on all three planets differ markedly from those predicted simply by the calculation of incident and reflected radiation. The atmospheres must hold the key to those differences and this impression is strengthened by the fact that Venus has a very thick, dense atmosphere and Mars a very thin one. If the sea-level atmospheric pressure on Earth is given a value of 1, the surface pressure on Venus is 90 and on Mars about 0.006. It is a large difference.

We might also look at the chemical compositions of the three atmospheres. On Venus the atmosphere is about 98 per cent carbon dioxide and on Mars about 95 per cent. The atmosphere of Earth contains about 0.03 per cent carbon dioxide.

As I explained earlier, solar radiation is most intense in the shorter wavelengths, especially those between 400 and 700 nm, that we recognize as visible light. On Earth the atmospheric gases are almost completely transparent to radiation at these wavelengths, the small amount of incident radiation that is absorbed being 'captured' mainly by particles, not gas molecules. Solar radiation warms the surface and the surface radiates heat, but at wavelengths from about 800 to 4,000 nm. The size of the carbon dioxide molecule (CO_2, or O—C—O) is such that it absorbs radiation with a wavelength greater than 1,000 nm and absorbs especially strongly between 1,200 and 1,800 nm. It also absorbs at about 500 nm, though this absorption of incoming radiation in a very narrow waveband is not very important.

When energy is absorbed the molecule radiates it again, but in all directions. Some goes upward, some sideways, and some downwards, and it may strike and be absorbed by other molecules. The overall effect is to trap radiation at these particular wavelengths and so to hinder its departure. Impart energy to a molecule and it responds by moving faster. This is the same thing as saying the substance of which the molecule is a part becomes warmer. When molecules strike our skin, the faster they move the more readily we detect them, and what our nerve endings detect we call warmth. So this trapping of outgoing radiation produces a warming of the atmosphere. The more carbon dioxide there is in the atmosphere the greater is the chance that photons will encounter a CO_2 molecule and so the greater will be the warming. The atmosphere of Venus contains a very large amount of carbon dioxide and therefore it traps a great deal of heat. The martian atmosphere also contains a large proportion of carbon dioxide but the atmosphere is too thin for the carbon dioxide to overcome the greater overall tendency of the martian atmosphere to transport heat upward by convection, making the planet even colder that it would be had it no atmosphere at all. Venus, on the other hand, has a very large atmosphere, and it is very hot.

As I explained in the first chapter, it is reasonable to

suppose that at one time composition of the atmospheres of Earth, Mars and Venus was similar and that Earth's atmosphere was also about 98 per cent carbon dioxide. Were that the case today, the trapping of outgoing heat by the atmosphere would give us an average temperature at the surface of the planet of between about 240 and 340°C (about 460 to 645°F).

It is the influence of certain gases, mainly carbon dioxide, on the surface temperature that is known as the greenhouse effect, although the name is not really appropriate because in a real greenhouse the warming is caused mainly by preventing air inside from mixing with air outside. On Mars the effect is limited because the atmosphere is very thin. On Venus, where the atmosphere is dense, it has run away to produce an extremely high surface temperature. Conditions are stable on both planets. From time to time storms on Mars, with hurricane-force winds, raise large amounts of dust into the atmosphere and this produces a warming effect, but it is temporary. The winds die down, the dust settles, and the temperature drops back to its former value.

Carbon Dioxide and the Earth's Climate

The effect on Earth has been much more interesting. Some four billion years ago the intensity of solar radiation was perhaps 30 per cent less than it is today. Conditions on Earth should have been distinctly cool. Had they been as cold as such a large reduction in the solar constant suggests there would have been little or no liquid water at the surface. The oceans would have been covered in ice, the lakes would have been ice sheets and the rivers glaciers. We can be very certain the Earth was not like that because rocks have been found that formed around that time, and they are sedimentary rocks.

Sedimentary rocks, as the name suggests, form when sediments are compressed beneath the weight of overlying rocks. The process takes millions of years. Sediments consist of material washed from rocks by moving water and deposited on the sea bed where they accumulate. Sedimentary rocks could not have been formed unless there were sediments earlier, and sediments cannot accumulate in the absence of running water and a liquid sea. The existence of those ancient rocks proves beyond doubt that four billion years ago there was a substantial amount of liquid water on the surface and that water had been present for millions of years before that - the time needed for the sediments to accumulate. This means that at least for much of the time the temperature must have been higher than the freezing point of fresh water (0°C, or 32°F).

The large amount of carbon dioxide in the early atmosphere caused a greenhouse warming that kept the temperature high enough for water to exist. In fact it may have been warmer on Earth then than it is today. Some scientists have suggested the average temperature was about 23°C (73°F), others that with perhaps 1,000 times more carbon dioxide in the air than there is now the temperature at the sea surface was close to 100°C (212°F). Certainly it was warm enough for life to become established. We know this because evidence of that life has been found in rocks about 3.8 billion years old. I should mention that modern techniques, based on the rate at which radioactive elements decay, are very sophisticated and allow the age of rocks to be determined in a number of ways that can be crosschecked against one another, so the dating is reliable to within a few million years - a degree of vagueness which is hardly important when the age being measured is counted in thousands of millions of years.

At the very beginning of life on Earth, therefore, the composition of the atmosphere produced a climate that was tolerable for living organisms. Since then, the burial of carbon, mainly in the form of carbonate (especially limestone) rocks has progressively reduced the amount of carbon dioxide in the atmosphere. This reduction has marched in step with the progressive increase in solar output with the result that the average temperature has remained between 10 and 20°C (50 to 68°F) throughout the entire history of the planet.

There have been fluctuations, of course, periods when over the world as a whole the climate was warmer or cooler than it is now, and carbon dioxide is not the only factor that influences the climate. The movement of continents also has an effect.

You can think of the outermost crust of the Earth as being

composed of rock that has cooled and solidified. This crust is a few tens of kilometres thick – thinner beneath the oceans than beneath the continents. The crust is not one solid unit but is composed of a number of sections that can move in relation to one another because they lie above a region in which the temperature and pressure are so high that although the material is extremely dense it is also plastic. It flows a little and convection currents occur within it. The plates that comprise the crust float on this denser material. By comparing rocks and the distribution of plants and animals it has been possible to reconstruct at least the recent history of such movements and so to trace changes in the Earth's geography.

The Wandering Continents

About 265 million years ago there was but one supercontinent, Pangaea, and one ocean, Panthalassa. Then, some 225 million years ago, Pangaea split into two smaller supercontinents, Gondwanaland and Laurasia. Later still these supercontinents split further. Gondwanaland became what are now South America, Africa, India, Australia and Antarctica. In time Laurasia became what are now North America, Greenland, and Eurasia (Europe and Asia). These continents moved, and are still moving. India collided with southern Asia, for example, and is still moving northward, pushing the Himalayas still higher as you might crumple a cloth by sliding it. The Atlantic Ocean did not exist at all 150 million years ago. Today it is still growing wider so Europe and America are drifting apart, and the Pacific is growing narrower.

The wandering of the continents exposes them to different climates at different episodes in their history. About 400 million years ago what are now England, Wales and Ireland are believed to have lain in the tropics south of the equator. By 300 million years ago Scotland had joined them and all four countries lay more or less on the equator. We have been moving northward ever since.

The movement of continents itself affects the global climate because of the change in the distribution of the Earth's albedo and because of the difference in the way land and sea warm and cool. Today there is a large land mass at the South Pole. Land stores heat much less efficiently than does water, so in very high latitudes it will be colder. The water on its surface will be fresh rather than salt and fresh water freezes at a higher temperature than salt water. Just the presence of such a large land mass in very high latitudes makes it more likely that there will be an ice sheet at the pole. Antarctica once lay in the tropics, of course, which is why it may conceal large reserves of coal, formed from the remains of plants that grew in tropical swamps. Coal measures in other countries that are now in middle or high latitudes - including Britain and most European countries, the USSR, China, and the USA - provide similar evidence of their passage through the tropics. In the same way, the large expanse of ocean that today lies at the equator produces the humid equatorial climate. When the vast area of Pangaea straddled the tropics the albedo would have been higher, there would have been less water, and the climate would have been quite different - for purely geographic reasons. The interior of Pangaea was probably a very dry desert, but deserts have always come and gone. The red sandstones that make south Devon so attractive and give its soils their red colour were once dunes in a large desert.

Such movement of continents seems to us, with our brief history, to occur very slowly, although it is quite rapid in the perspective of the history of the planet. By a curious coincidence, the continents move at just about the same rate as human toenails grow.

Within the climatic regimes which the continents experience there are other changes, with shorter time spans. The most dramatic, and in terms of our worries about the greenhouse effect the most important, are ice ages and the interglacials that punctuate them. These are relevant because they are linked closely to the greenhouse effect.

Ice Ages

There have been ice ages at intervals throughout most of the Earth's history. There is evidence of ice ages that occurred more than two billion years ago, Laurasia and Gondwanaland experienced at least 15 major groups of ice ages, and there have been many ice ages since. What are now central and southern Africa have been covered by ice more often than any other part of the world. At present we are living in an interglacial, a brief warm spell between two ice ages, and during the last two million years or so the ice sheets have advanced and retreated repeatedly. It seems that it is normal for the Earth to have some areas covered with ice and that we have now reached a period in the history of our planet when we should regard an ice age as the usual state of affairs. Today about 10 per cent of the Earth's surface is covered by permanent ice. At the peak of the most extensive of the recent ice ages the figure was about 28 per cent.

An ice age can begin quite rapidly, taking the climate from comparatively warm conditions to full glaciation in a few centuries or even less. This is a dramatic event but it does not represent a very large change in the average temperature. During the most intense ice age the average temperature is only about 7 or 8°C (13 or 14°F) cooler than it is now. A quite small drop in temperature can initiate an ice age, and a small rise can bring one to an end.

This is another example of a system that is extremely sensitive to small differences in its starting conditions. Imagine what would happen were the summer temperature to fall just a little. The winter snow would arrive as usual in the middle and high latitudes. As spring arrived the thaw would begin but because the summer temperature was somewhat lower the thaw would be slower and the boundary between the snow that remains all year round and the snow that melts in summer would be extended a little. There would be a rather larger area of year-round snow. Probably only the scientists who measure such things would notice the difference and even they might not be sure about it.

This slightly larger area of snow would increase the albedo so more sunlight would be reflected. This would add to the cooling. Next winter more snow would arrive and the following summer the area of year-round snow would extend again, a little further, and the albedo would increase. This is positive feedback, a response to change that tends to accelerate the rate of change rather than acting to reverse it. Each year the area of year-round snow would increase, that increase would make the temperature fall, and the rate of cooling would increase from year to year.

The ending of an ice age is also subject to positive feedback. As ice melts the albedo is reduced and the surface absorbs more warmth, so accelerating the warming.

These are the kind of changes climatologists fear we may experience if we induce a change, in either direction, in the average temperature of the planet. If we cool the planet just a little we could plunge ourselves rapidly into an ice age. If we warm the planet by just a few degrees we could trigger the melting of the polar ice caps. Once it began, such a change, driven by positive feedback, might be impossible to reverse. We would just have to live with it.

We now have quite a lot of information about what happens to the climate during ice ages because it is recorded for those who know how to read the record. When you make ice in a refrigerator you pour water into a tray and freeze it. That is not the way ice sheets form. They begin as snow. The snow falls, as snowflakes or crystals, and once fallen it does not melt. More snow falls on top of it, and more on top of that, and little by little the weight of the overlying snow compacts the lower layers until they are converted from flakes and individual crystals into solid ice.

The snow at the top is quite loose. This means there is air between the crystals. As the snow is packed down and turned into ice some of this air is squeezed out but not all of it. Tiny pockets of air remain trapped and they are the air that formed part of the atmosphere at the time the snow fell. They have the chemical composition of that air. In the great ice sheets of Greenland and Antarctica this accumulation of ice and trapped, or entrained, air has been going on year after year for perhaps two million years.

During the 1980s, a team of French and Soviet scientists at the Soviet Vostok station in Antarctica drilled deep into the ice sheet, extracted cores from it, and then studied the air and water they obtained from the cores. One part of their study concerned the water, H_2O. The nucleus of an atom is made up of protons and neutrons. If the number of protons increases or decreases the physical and chemical properties of the nucleus change and it becomes a nucleus of a different element. If the number of neutrons increases or decreases, however, the mass of the nucleus changes but not its properties. It remains the same element, but a different isotope of it. Most elements exist in several isotopic forms and there are two useful isotopes of oxygen,

O-16 and O-18, and two of hydrogen, H-1 and H-2 (which is also called deuterium). The proportions of these four isotopes in the atmosphere never change, but they do change when they combine to form water and that change depends on the temperature. The colder it is the more O-18 and H-2 the water contains. So if you can analyse the water and measure the proportions of these isotopes in comparison with the proportions in the air you can calculate the temperature at which the water formed. The isotope ratio is a kind of thermometer. By performing these measurements and calculations for the water from the ice cores the scientists were able to tell the temperature, and also the difference between summer and winter temperatures. This allowed them to measure the seasons and so to date the cores rather as you might tell the age of a tree by counting the growth rings in a cross-section of its trunk.

Then they investigated the composition of the entrained air. It contained carbon dioxide, of course, and when they compared the amount of carbon dioxide with the temperature at which the water formed the two sets of figures followed one another so closely that they were able to relate even very small fluctuations in the carbon dioxide concentration to small temperature fluctuations. The amount of carbon dioxide in the air and the temperature were locked together. More carbon dioxide meant warmer conditions, less carbon dioxide meant cooler conditions, and the onset and end of ice ages themselves was associated with changes in the amount of carbon dioxide.

Is it a change in the concentration of carbon dioxide that initiates and ends the ice ages, or does something else bring about the climatic events, so the change in carbon dioxide concentration is an effect rather than a cause? At present no one can say with certainty, although it is clear that once a climate change begins, for whatever reason, it is taken over and driven further by alterations in the amount of carbon dioxide in the air. Nor can there be the slightest doubt that should the concentration of carbon dioxide be altered a change in the climate will follow.

The Effects of an Ice Age

The prospect of an ice age may strike us as catastrophic, but this is because we live in a latitude that would be affected seriously. Much of Britain and the northern part of North America have been covered by ice many times, and beyond the edges of the ice sheets there was permanently frozen ground not far below the surface (permafrost), and sparse vegetation that was typical of the tundra regions you find today in the far north of Canada and Europe. This makes it sound an uncomfortable place to live but the people who lived during recent ice ages felt differently about it. They lived quite close to the edge of the ice sheet, probably because that was where they found large herds of animals they could hunt for food, skins to make warm clothing, and bones and antlers to make tools. In the world as a whole, however, an ice age is much less spectacular. The temperature in the tropics hardly changes and although the ice may cover as much as 30 per cent of the Earth's surface this still leaves 70 per cent icefree.

An ice age does have important consequences, even so. The accumulation of snow, year after year, steadily removes water from the oceans. In places the Greenland ice sheet is more than 3 km (nearly 10,000 feet) thick. This is a great deal of water and during an ice age much more is trapped. The depletion of the oceans causes the sea level to fall throughout the world and this exposes more land. So the fact that a large area is covered by ice does not mean there is less space for land plants and animals to live, provided they are prepared to move. Moving itself is easier, too, because land bridges are exposed. The first humans entered North America from Asia across the land bridge that linked Alaska to Siberia across the Bering Strait and it is not so long since humans walked dry-shod between England and France.

When the ice sheets melt and the glaciers retreat, sea levels rise again as water is returned to the oceans. The warming of the oceans also causes sea level to rise because the water expands. The sea-level changes do not take the form of a steady progression or regression, however. They rise and fall in stages punctuated by intervals in which nothing much seems to be happening. When the rise or fall comes it can be sudden.

This is obvious if you think about the way it occurs. For years on end the rising tide in a particular place is checked by high ground to the seaward side of lower ground. Then, one year, a high spring tide is accompanied by a storm surge caused by several days of strong onshore winds out at sea. The sea flows over the high ground, washing some of it away, and inundates the low ground. Once that happens the barrier has been removed and the coastline has shifted inland.

Where it forms on land, the immense weight of an ice sheet causes the ground beneath it to sink. This has no effect on sea level around the affected region because the depressed areas lie beneath a great thickness of ice, but when the ice melts the depressed land slowly rises again. At present Scandinavia and Scotland are still rising, so they stand higher than they did in relation to the sea, but south-eastern England is sinking as a result as Britain tilts like a see-saw whose fulcrum is somewhere in the Midlands.

This process is called isostatic adjustment, based on the concept of isostasy. According to this concept a plate of the Earth's crust floats in equilibrium on the material beneath it. If the mass increases in one part of the plate, that part will sink but the sinking must be compensated by the rising of another part somewhere else. The ice increases the mass of the crust on which it lies and so that part sinks and somewhere else another part rises - and when the ice melts the process goes into reverse.

This means it is not at all a simple matter to calculate the long term changes in sea level that follow the formation or melting of ice. The short term change is rather easier to understand because it involves only the water. The rising or sinking of the land itself is a much slower process. Add water to the oceans and the sea level will rise, remove it and the sea level will fall, and the extent of the change can be calculated. Then, after the short term change has happened the land itself will start to move, rising in some places, sinking in others, until a new sea level is established.

We also have some knowledge of past sea levels and can use this to map the location of past coastlines. During the most extensive of the recent ice ages, for example, the sea level was 130 metres (425 feet) lower than it is today. The English Channel, Western Approaches, Irish Sea and North Sea were all dry land. The Scillies, Orkneys and Shetlands were ranges of hills. Were the present polar ice caps to melt completely, the sea level would rise perhaps by as much as 80 metres (260 feet), reducing Britain to a group of islands. In this context it is the fate of the Antarctic ice that matters.

Ice that floats on the sea simply returns to the sea without increasing its volume. It is the ice on land that represents water which has been removed from the sea. There is no land at the North Pole, and Greenland is the only large land area covered by ice in the northern hemisphere. Antarctica, on the other hand, is a land mass twice the size of Australia. Not all of it is covered by ice, but a considerable proportion is. So it is the Antarctic ice sheet the scientists watch most carefully.

At present the sea level is rising slowly due partly to the warming of the sea and consequent expansion of water, and in some places due partly to the isostatic adjustment that follows the ending of the last ice age about 10,000 years ago.

Ice ages may begin rapidly, but they do not necessarily end smoothly. It is not a matter of a uniform and sustained rise in temperature. At the end of the last ice age temperatures began to rise quite quickly, but then the warming was halted abruptly and there was a cold phase. That ended, according to recent evidence very abruptly indeed - in a matter of a few decades - and the warming resumed. Then it halted for a second time and there was a second cold phase. These two episodes are called the Older and Younger Dryas, because a small, cold-climate plant called *Dryas octopetala* (mountain avens) grew widely in places that are now too warm for it to occur naturally. No one knows why the warming was reversed but there are theories. One of these holds that the rapid warming caused an equally rapid melting of the ice and the melted ice flooded huge amounts of fresh water into the sea over a short period. Fresh water is less dense than sea water, the two do not mix very readily, and the fresh water floated on the surface, eventually drifting over a large area. Then, presumably during a hard winter, the fresh water froze and that large area of sea came to be covered by ice. This chilled the air moving across it, which cooled the land, and so triggered the cold spell that was intensified by its own positive feedback.

Even within ice ages and interglacials there are smaller fluctuations that last for a few centuries. The Little Ice Age is one such fluctuation. It amounted to only a small decrease in the average temperature, no more than 1 °C at its most extreme, and the evidence that it occurred at all is not based on direct records of temperature, but on much more accurate records of such things as harvests, the width of annual growth rings in trees, and on the advance and subsequent retreat of European glaciers. Some historical records are reliable because they were important at the time. The date of the first frosts, for example, were of great importance in some wine-growing areas, and the date the river Neva froze over at Leningrad was important for trade.

The Little Ice Age began some time in the middle of the sixteenth century, and continued until the early nineteenth; it affected all of the middle latitudes of the northern hemisphere, with evidence for it in North America and Asia as well as in Europe. For most of that time people had no knowledge of the great ice ages of the past but had they known of them they might well have come to believe their lands were soon to vanish beneath the advancing ice sheets. Then, in the nineteenth century, conditions began to grow warmer and the warming continued until the middle of this century, when it gave way to a cooling and scientists really did start predicting at least another Little Ice Age if not a full-blown one. Since then the warming has resumed.

The Little Ice Age may or may not have been caused by the change in solar activity marked by the Maunder sunspot minimum, but it is believed that during it the North Atlantic Drift altered its course and flowed further north so it no longer washed the shores of north-western Europe. This would have led to a cooling of the air flowing from the Atlantic over north-western Europe and as the air masses continued in their eastward progress it might have contributed to the cooling in central Europe.

The melting of sea ice may contribute directly to the concentration of carbon dioxide in the atmosphere, thus accelerating the warming. Carbon dioxide is slightly soluble in water and so it dissolves into the sea. When sea water freezes, the ice consists exclusively of fresh water. As the water molecules move more slowly, then move closer together to form crystals, it is only the constituent atoms of hydrogen and oxygen that are involved. You can check this for yourself easily enough. Dissolve about 3.5 parts of table salt in 100 parts of tap water and the resulting solution will be as saline as average sea water. Place the solution in the freezer, wait for it to freeze, then lift the ice from the surface of the liquid beneath. If you taste the ice you will find it contains no salt, but the remaining liquid is much saltier than it was before you froze it.

This is the start of what happens when the sea freezes. The water just below the ice contains the salt from the water that has frozen and so it is more saline than the water below it. Being more saline it is also more dense, and so it sinks, eventually to a considerable depth, and its place is taken by lighter, less saline water. The dense brine also contains dissolved carbon dioxide and so this, too, is carried deep below the surface and the water that replaces it is unsaturated so more carbon dioxide can dissolve into it. This mixing is a very effective way of removing carbon dioxide from the air, but when the ice melts the situation changes. Now fresh melt water, less dense than salt water, floats on the surface and mixes only slowly and then only with the surface layer of the sea. It is soon saturated with carbon dioxide and the 'scavenging' mechanism ceases to operate. If carbon dioxide is being added to the air while this is happening it will accumulate there more rapidly because the sea is removing it more slowly.

Despite the huge amount of water lying on its surface, Antarctica is one of the driest deserts in the world. Fresh snow does fall, of course, but the annual amount is small. In the interior of the continent, at latitudes higher than 75°S, the annual precipitation, converted to its equivalent as rainfall, is no more than 41 millimetres (1.6 inches) and close to the South Pole itself it is about one-quarter of that. The violent blizzards for which Antarctica is notorious consist of snow blown up from the surface (indeed, technically that is what distinguishes a blizzard from a snowstorm). For comparison, annual rainfall in the Sahara averages about 51 millimetres (2 inches) and in the driest, central areas it is half of that. Antarctica, then, is more arid than the Sahara.

Dry conditions are typical of ice ages, as you might expect. Cooler air temperatures mean less water evaporates and an overall reduction in sea area means a smaller surface area of water is available from which water can evaporate. By and large it is during ice ages that deserts form. During the warmer episodes that punctuate the ice ages, the interglacials, the opposite holds true and these are usually times of wet climates when deserts disappear. As I shall explain later, some of the present greenhouse predictions anticipate increasing aridity in the centres of the continents. These predictions could be wrong and even if they are correct the aridity might be temporary, giving way to wetter conditions as the warming continues.

We live in an interglacial, one of these interludes of warmth between much longer cold periods. During the present cylce of ice ages interglacials have lasted an average of about 10,000 years. Our interglacial, the Flandrian, began about 10,000 years ago and other things being equal we must be very close to its end. Of course, very close can mean anything from within a few years to within about 3,000 years, but when it arrives the ice is likely to arrive quickly. Given that aridity is typical of cold periods, increasing drought in such semi-arid regions as the southern border of the Sahara (the Sahel) and the spread of deserts, which according to some people is happening everywhere, might suggest that an ice age is imminent, but only if the reports are what they seem. Are the deserts really spreading, or is bad land management causing erosion that is interpreted as desertification? Are there more droughts, or do droughts tend to occur in bunches at fairly regular intervals so no special meaning can be read into those of recent years?

The questions heavily outnumber the answers. The history of the world's climates, and especially their recent history, suggests we may expect an ice age to begin very soon. If in fact a greenhouse warming is occurring, or is about to occur, we should see it in this context, as a phenomenon that is overriding the natural climatic trend.

If there is to be a greenhouse climate it must be because directly or indirectly we - the humans - are releasing into the atmosphere gases that trap outgoing radiation. These are the greenhouse gases. Carbon dioxide is one of the greenhouse gases, but only one of them. There are others and we need now to consider each of them in turn to see whether their atmospheric concentration is changing, and what effect each may have.

CHAPTER 5

The Greenhouse Gases

Some years ago, James Lovelock and I wrote a book called *The Greening of Mars*. It was a kind of scientific speculation, somewhere between a novel and a book about science - one reviewer called it a prospectus. Our central theme was the transformation of the martian climate in such a way as to make that cold, dry planet habitable for species from Earth, including humans.

The plan was quite simple, at least in principle. Later, two groups of scientists, working independently, tested it. They made computer models of the real martian climate, did to the atmosphere the things we described in our book, and right enough the climate improved. In our scheme we began with some chemicals that were surplus to requirements on Earth, and some ballistic missiles we filched as they became obsolete from the defence programmes of a few countries. We strapped rockets together to give them more power, made the chemicals into payloads that took the place of the warheads, and fired our missiles at Mars. They were required only to hit the planet, burst into fragments in impact, and disperse their chemicals into the air.

Once airborne on Mars, the chemicals started trapping heat radiated from the martian surface, so warming the atmosphere. As the atmosphere warmed, carbon dioxide, present at the martian poles as a dusting of ice, sublimed into the air (carbon dioxide does not exist naturally as a liquid - it changes phase directly between the solid and the gas). This made the atmosphere warm still more, so water ice also began to sublime - which it would do under the very low martian air pressure. We did other things, too, to alter the planetary albedo, but essentially we were introducing one substance that is not present on Mars and using it and other substances that do exist on Mars to produce a large and rapid greenhouse effect. By the time we had finished with the place, green plants were growing in the open, there were marshy areas and shallow lakes, and in the daytime you could have walked around on Mars in shirt sleeves, provided you remembered to take your breathing apparatus for the air still contained very little oxygen. You had to be home before nightfall, however, for the temperature still dropped sharply after sunset.

Trapping Radiation

The chemicals, which I said were surplus to requirements on Earth, were chlorofluorocarbons, or 'freons' to use the name given them by Du Pont, the company that invented them. Compounds of chlorine (Cl), fluorine (F), and carbon (C), they are better known these days as CFCs and belong on the long, long list of chemicals we love to hate. In our book we assumed, reasonably enough, that, under pressure from environmentalists, in due course their use would be banned, leaving quite large stockpiles of them that could not be disposed of easily because the important thing was to isolate them from the atmosphere. At least that is what people believed. We were doing the chemical industry a favour.

We used CFCs rather than, say, carbon dioxide because, although both are greenhouse gases, CFCs are at least a thousand and perhaps up to 20 thousand times more effective than carbon dioxide at trapping long wave radiation. A little of them goes a long way!

So what does trap long-wave radiation? For once the answer is simple. More or less any molecule will do so if it comprises three or more atoms, because then its size will be comparable to the wavelength of some of the radiation. Because of this, if it intercepts radiation it will move, spin, or vibrate more rapidly - it will become hotter - and will impart some of its newly acquired energy to neighbouring molecules.

Beyond that, each different size of molecule absorbs at different wavelengths. The efficiency of a greenhouse gas depends on its concentration in the atmosphere and also on the waveband in which it absorbs. The Earth's long wave radiation peaks in particular wavebands, most strongly at wavelengths of around 1,000 and 1,200 nm, so molecules that absorb at these wavelengths will have a larger effect than molecules that absorb at most other wavelengths. If a substance that absorbs at a particular wavelength is present in amounts large enough to have a significant effect, the addition to the atmosphere of another substance absorbing at the same wavelength will make little difference - just as adding a cupful of water to the bath will have little effect if the bath is already overflowing.

Whether or not a molecule will intercept radiation, and whether the radiation will accumulate, also depends on the length of time it remains in the atmosphere. The air has a great capacity for cleaning itself. As I mentioned earlier, oxygen and hydroxyl are very active chemically and will oxidize many substances. Once oxidized, a substance is usually less reactive and if it is a pollutant this means it will be less harmful. Oxidation also increases the size of molecules by combining them and larger molecules are more likely to be washed to the ground by the rain. So molecules remain in the air for different lengths of time. On average, a molecule of nitrous oxide remains airborne for 170 years, one of carbon dioxide for 100 years and one of methane for 10 years. CFC molecules remain airborne for several decades.

Carbon dioxide is the best known greenhouse gas because it is by far the most abundant and it absorbs at peak wavelengths. If this absorption causes a warming, more water will evaporate. Water vapour (H_2O) has three atoms to its molecule and it, too, is a greenhouse gas. It absorbs at about 700 to 900 nm and so it blocks a different radiation window from those blocked by carbon dioxide.

Methane (CH_4) has five atoms in its molecule and its absorption overlaps that of water vapour. In still other parts of the radiation spectrum there is absorption by nitrous oxide (N_2O) and ozone (O_3).

Ozone and Ultra-Violet Radiation

Ozone is particularly interesting because it also absorbs incoming short wave radiation. This has little effect in the troposphere but is quite important in the stratosphere. That is where the ozone layer is found, a place between about 10 and 50 km (6 to 30 miles) above the surface, where ozone is concentrated, most of it in the middle of the layer, between about 15 and 30 km (9 to 19 miles). The term 'ozone layer' makes it sound as though the ozone is very common in this belt, but in fact its concentration is only from about 1 to 10 parts of ozone to every million parts of other gases. Even in the ozone layer, where most of the planet's ozone is to be found, ozone is a very rare gas.

It is a form of oxygen. Ordinary oxygen exists as O_2, a molecule made of two atoms of oxygen bound together. Just beyond the spectrum of visible light, at wavelengths between 4 and 400 nm, the radiation we receive from the Sun is called ultraviolet, or UV for short. You cannot see it but, having a short wavelength, it has more energy than visible light and the more energetic end of the UV waveband, with wavelengths shorter than about 300 nm, can damage living tissue.

This damage should not be exaggerated. Heavy and prolonged exposure can cause a form of skin cancer in humans - the cancer is easily diagnosed and can be cured - but there is little other effect and UV cannot penetrate below the surface layers of human skin. The reported increase in skin cancer in recent years probably results from the fashion for prolonged sunbathing by fair-skinned people who are especially susceptible to burning because they have fewer of the skin cells that make melanin than dark-skinned people. Melanin is the dark pigment that gives you a suntan, and it absorbs UV. A suntan is visible evidence that the body has defended itself against injury and is as much an indication of robust good health as is scar tissue.

Plant cells can also be damaged by UV, but it cannot penetrate water so readily as visible light does. I shall explain in a moment what the notorious ozone hole is, but beneath the hole, where exposure to UV was greatly increased, scientists made measurements in October and December, 1988, of the effect on the very abundant phytoplankton found in Antarctic waters. At the time they estimated that the intensity of UV radiation was double the intensity found during a normal summer. There was a small reduction in phytoplankton, but it was confined to the surface layers. The intensity of UV was reduced by 25 per cent in the uppermost metre of water and there were no effects at all at any depth greater than 20 metres (65 feet), where these tiny, single-celled plants were unaffected. Overall, the reduction in photosynthesis amounted to only a few per cent. So although there are dangers arising from over-exposure to UV they are much less dramatic than people have been led to fear.

When oxygen molecules are exposed to radiation at the higher end of the UV waveband the energy imparted to the molecules is sufficient to break the bond that holds them together. Molecules of oxygen (O_2) become single oxygen atoms (O). This is not a stable condition for oxygen and the atoms soon recombine, but instead of combining with other single atoms to reform ordinary molecules, some of them combine with molecules, so $O_2 + O \rightarrow O_3$, and O_3 is ozone. The ozone is also exposed to UV, which breaks its bonds, so the three atoms separate and rejoin. The process continues for as long as the air is exposed to the UV radiation that drives it and because that radiation provides the energy for the process the energy itself is absorbed. This is how stratospheric ozone absorbs the short wave part of the UV, thus preventing it from penetrating to the surface of the Earth. The depth of air through which the radiation must pass before it strikes sufficient oxygen molecules for significant amounts of ozone to form leads to the concentration of ozone in the region of the middle stratosphere called the ozone layer. At higher levels the air is so rarefied that UV affects few molecules as it passes through.

Because the process depends on the intensity of UV, which is part of the solar radiation we receive, the amount of ozone in the stratosphere changes according to the intensity of sunlight. This varies widely from place to place, from season to season, and even from day to day. The ozone layer is thickest over the Arctic and Antarctic during their summers, when the days are very long, and thinnest during their winters, when there is little daylight. In polar regions the formation of ozone is amplified in summer because although the Sun sets for only a brief period it never rises very high in the sky so its radiation passes through the atmosphere obliquely. This means the radiation travels a much greater distance through the atmosphere and therefore is more likely to strike oxygen and ozone molecules. The ozone layer is also thicker during summer than in winter in mid latitudes, for the same reason.

Over the equator there is no seasonal change because there are no seasons, and UV penetration is greater because solar intensity is greater and the atmosphere is effectively thinner because the Sun is directly overhead and radiation reaches the surface vertically, travelling the shortest possible distance. If the ozone layer were to disappear completely over the middle latitudes - say over Britain - the amount of UV reaching the surface would be about the same as ordinarily reaches the surface at the equator. If the consequences were as calamitous as some people suggest you would expect life to be difficult if not impossible at the equator. Clearly it is not.

The Stratosphere

The stratosphere contains trace amounts of many gases and these are involved in a large number of reactions. The chemistry of this region of the atmosphere is extremely complicated and difficult to study because of its remoteness. What is known, however, is that some of these reactions can lead to the formation of stable compounds that remove ozone, and chlorine is the element most seriously implicated in such reactions. Even then, the conditions required for chlorine to deplete ozone are very unusual. The ozone hole is a curious phenomenon.

During the Antarctic winter the air in the stratosphere becomes extremely cold. The air movements around the pole create at their centre a vortex, a region of still air much like the eye of a storm. Within this vortex thin clouds of ice crystals can form, called polar stratospheric clouds. The reactions in which chlorine removes ozone involve nitrogen compounds and they occur on the surfaces of these ice crystals just as the winter is ending and the sunlight supplies a little energy. In some years the depletion has been severe at some altitudes and the total depletion amounted to 50 per cent in September, 1987, although it was only 15 per cent the following year. The depletion was linked to the presence of chlorine monoxide (ClO), the compound that prevents the oxygen atom from joining an oxygen molecule to form ozone. Later, studies over the Arctic suggested that that region, too, may have experienced a small reduction, of about three per cent, in the ozone layer in at least one year. Depletion is less likely in the Arctic because stratospheric temperatures are higher than those in the Antarctic and it is more unusual for a circumpolar vortex to form.

At one time oxides of nitrogen were suspected of depleting ozone and it was feared that large fleets of airliners flying in the stratosphere could cause damage because of the nitrogen oxides in their exhaust gases. In those days it seemed possible that in addition to the Anglo-French Concorde there would also be fleets of US and Soviet supersonic airliners operating at these altitudes. In the event the fleets of high-flying airliners were not built and then it was discovered that among the reactions in which nitrogen oxides engage there is one with chlorine whose end product is chlorine nitrate ($ClNO_3$), a stable compound that prevents the chlorine from reacting with oxygen and ozone. The nitrogen oxides would have helped protect the ozone rather than depleting it.

Although the Antarctic hole was very large at times and extended over some inhabited areas, it occurs for only a few weeks and there is no firm evidence of stratospheric ozone depletion except in these very high latitudes, although there may have been depletions of between 2 and 10 per cent in winter and early spring in the middle and high latitudes of the northern hemisphere. No one knows whether these are normal seasonal fluctuations or whether they result from human activities.

Measurements in North America suggest that in fact the amount of UV reaching the surface in most parts of the world has been decreasing in recent years, partly because there has been an increase in particulate matter in the stratosphere and the particles scatter UV, and partly because ozone is now a pollutant in the troposphere, so the total amount of ozone in the air is increasing, especially in and near to industrial areas. Ozone absorbs UV wherever it occurs. It does not need to be in the stratosphere to perform this function.

The polar stratospheric clouds have been found to contain particles, possibly of nitric trihydrate, at just the heights at which ozone depletion occurs but there is something of a mystery about the clouds themselves. Sometimes clouds much like them are visible, as noctilucent clouds, between about latitudes 50 and 60° in either hemisphere. As the name suggests, they are seen at night, in summer, and they are very beautiful. Blue or yellow in colour and wispy in appearance, they often have a wave-like formation and move slowly – in fact they are travelling at up to 550 km per hour (345 m.p.h.), but at a height of 80 to 85 km (50 to 53 miles) so their movement seems slow.

The mystery concerns the source of the water to make them, for virtually no water or its vapour can cross the tropopause into the stratosphere. One theory is that very small amounts of methane are carried across the tropopause. In the presence of oxygen, methane breaks down and one of its products is water which can then freeze on to dust particles from meteors. Over the past century the atmospheric concentration of methane has increased from about 0.9 parts per million to 1.7 parts per million and it is now increasing rapidly. There is also a possibility that noctilucent clouds are more common than they were, and perhaps that they were extremely rare or even did not occur at all more than a century or so ago. So if we are to consider ways of preventing such depletion of the ozone layer as may be due to human activity we might do well to remember the effect methane may be having.

The source of the chlorine necessary for ozone depletion is generally supposed to be CFCs, but this has not been proved and there are other candidates. The eruption of the Mexican volcano El Chichón in 1982 injected a huge amount of dust and gases into the atmosphere, some of it into the stratosphere. This material may have supplied chlorine for ozone depletion. If so, the ozone hole should disappear in the next few years and already there are signs that the depletion may be less severe than it was. Various other industrial chemicals contain chlorine. There is carbon tetrachloride (CCl_4), for example, a common cleaning fluid whose atmospheric concentration is increasing by 1 per cent a year, methyl chloroform which is used in typing correction fluids and whose atmospheric concentration is increasing at 7 per cent a year, and chloromethane (CH_3Cl) which has many uses. Chloromethane also occurs naturally. Some species of wood-decaying fungi emit it, for example, at an estimated five million tonnes a year – compared with a world-wide release into the air of about 26,000 tonnes a year of CFCs. Perhaps we should not be too hasty in our rush to condemn the CFCs.

CFCs may not be guilty of depleting the ozone layer, and such a depletion may be less serious than is often supposed, but they remain effective greenhouse gases. This alone justifies restrictions in their production and use provided safe and satisfactory alternatives can be found for them. The main advantage of the CFCs is that they are very safe to use. They will not burn – some versions of them are used in fire extinguishers – and they are not poisonous to any living organisms. They replaced such compounds as pressurized carbon dioxide which will freeze flesh on contact, ammonia which is poisonous, and liquefied petroleum products which are highly flammable, and all of which are greenhouse gases. CFCs can be modified to make them break down before they reach the stratosphere, for example by adding a hydrogen atom to the molecule, but this may make them poisonous and the new products will still absorb long wave radiation.

The absorption of UV in the stratosphere means less solar energy reaches the ground. The absorption warms the stratosphere and cools the troposphere. Were the ozone layer to be substantially depleted this would contribute to climatic warming.

In the lower atmosphere, ozone is a greenhouse gas itself, though a minor one. Short-wave radiation that drives the reactions leading to ozone formation works as well close to the ground as at higher levels under certain circumstances. When the air contains hydrocarbons – compounds consisting mainly of carbon and hydrogen – and oxides of nitrogen, and the sunlight is very bright, photochemical reactions take place that produce smog, one of whose constituents is ozone. Ozone also forms when oxygen is exposed to large amounts of energy. Lightning flashes produce ozone and so do ordinary electrical sparks. You can sometimes detect a very characteristic, rather acrid smell immediately after a spark has flashed. The smell is of ozone. In quite low concentrations ozone causes respiratory irritation and at rather higher concentrations it damages respiratory tissue. Years ago people used to believe there was ozone in sea air, that this gave sea air a characteristic smell, and that it was good for your health. In fact there is less ozone in sea air than in air over land, ozone is harmful, and the health-giving smell people associate with the seaside is actually caused mainly by rotting seaweed. I have no idea whether rotting seaweed is good for your health but I suspect it is not!

From time to time the amount of ozone in the air has been measured so we know that about a century ago at one measuring point in Europe the air contained about ten parts of it per billion parts of other gases. Today in much of western Europe ozone levels near the ground surface are often 20 to 40 parts per billion and sometimes higher, and they can reach 100 parts per billion there and also in parts of the USA and in industrial areas of Australia. The hydrocarbons and nitrogen oxides needed for the formation of smog come mainly from vehicle exhausts but not exclusively so. The burning of surface vegetation also releases them.

In terms of their contribution to climatic warming, the major greenhouse gases are carbon dioxide (50 per cent), methane (18 per cent), CFCs (14 per cent), tropospheric ozone (12 per cent), and nitrous oxide (6 per cent).

Carbon Dioxide

We release carbon dioxide whenever we burn anything containing carbon. What we know as burning in fact is the rapid oxidation of carbon to carbon monoxide and carbon dioxide (in the air the monoxide then oxidizes further to dioxide), a reaction that emits heat. All our common fuels - wood, peat, coal, natural gas, and oil - are based on carbon and there is no way they can supply energy without emitting carbon dioxide because the combining of oxygen and carbon to produce carbon dioxide is the reaction by which they yield energy.

Obviously, the more carbon-based fuel we burn the more carbon dioxide we release into the air. In 1985, the biggest contribution, of 1,200 million tonnes, came from the USA, 35 per cent of it from electricity generation and more than 25 per cent from vehicle engines. Japan emitted rather more than 200 million tonnes, Britain, France and West Germany together rather less than 200 million tonnes, and the whole of Latin America 280 million tonnes. Such a comparison of countries and regions is not quite fair because their populations are not all the same size. If you measure carbon dioxide output per person, East Germany produces 5.4 tonnes, the USA slightly less than that, Britain 2.8 tonnes,

France 1.7 tonnes, and West Germany 2.9 tonnes. French and West German emissions fell between 1986 and 1987, by 3 and 2 per cent respectively, because in these countries energy was used more efficiently and in France nuclear power made a larger contribution to energy supplies than it does in most other countries. Nuclear power generation releases no carbon dioxide. In 1987, emissions were growing fastest in the USA, USSR, and China. The world total emission in 1987 reached 5.6 billion tonnes of carbon a year, an increase of 1.6 per cent since 1986 and of 10 per cent since 1983. At present the annual rate of increase world-wide is probably about 1.5 per cent. There is no doubt that in one way and another, but mainly through the burning of carbon-based fuels, the amount of carbon dioxide in the air is increasing. In a century it has grown from about 290 parts per million to its present 350 parts per million.

This sounds very little but you must remember that the growth is exponential – it works like the compound interest on your bank overdraft. At a growth rate of 1.5 per cent a year the amount (or the overdraft) will double in about forty six years.

Some of the carbon dioxide humans release is the result of forest clearance. There is disagreement about the size of this contribution but it is certainly large, probably amounting to more than one billion tonnes a year, and the emission occurs in two ways. If forests are cleared to provide land, usually for some form of agriculture, much of the vegetation that is removed has no use. Timber will be sold, of course, but the remainder is usually burned. The vegetation is also burned when poor grassland is cleared. The burning may provide land for ploughing or, left to regenerate, more nutritious grasses often replace the vegetation that is destroyed. At present, forests are being cleared at something like 104,000 square km (40,000 square miles) a year. To give you an idea of what an area that size is like, the area of England is just over 130,000 square kilometres (50,000 square miles).

Even without burning, there is a risk that forest clearance will liberate stores of carbon dioxide held in the soil. When living organisms, including plants, die, they are decomposed at or just below the soil surface. Tree roots, of course, decompose entirely below the surface. The decomposition also involves oxidizing the carbon this biological material contains, and so carbon dioxide is produced. Much of the carbon dioxide produced below the surface is held in the innumerable small pockets of air between soil particles, so enriching the soil air in carbon dioxide. The carbon dioxide content of the air contained in the top metre or so of a given surface area of fertile soil is many times greater than the carbon dioxide contained in the column of air above that area, from ground level all the way to the top of the atmosphere. While the area remains covered by vegetation the carbon dioxide is recycled constantly. Should the area be cleared, however, it is possible that the soil 'reservoir' might empty suddenly, releasing its carbon dioxide in a big surge.

The quantity of carbon dioxide emitted, however, is much larger than the change in atmospheric concentration would suggest. Only part of the carbon dioxide we release accumulates in the air. The remainder is absorbed, mainly by the oceans. Until recently it was believed that the oceans absorb half of all the carbon dioxide we release but more recent research suggests the proportion is much smaller, around 30 per cent.

The oceans both absorb and release carbon dioxide, most of which is cycled naturally, as a result of such ordinary processes as photosynthesis, respiration, and the decomposition of wastes. It is estimated that each year the oceans absorb 105 billion tonnes and release 102 billion tonnes, so that overall they remove three billion tonnes a year from the atmosphere (although, as mentioned above, this figure may have to be revised). Were they not absorbing this amount of carbon dioxide the gas would be accumulating in the air at about twice the rate it has been accumulating over the past half century or so. At the same time, were humans not pumping carbon dioxide into the air it is at least possible that atmospheric carbon dioxide levels would be falling and the global climate would be heading, perhaps rapidly, towards a new ice age.

Carbon dioxide is also used by marine organisms for photosynthesis, it is released by them as a by-product of respiration, and carbon is passed from one organism to another as marine plants are eaten by animals and small animals are eaten by larger ones. Much of the debris from this biological activity coagulates to form pellets or lumps that sink fairly rapidly and about one-tenth of this material sinks to below the thermocline - the oceanic equivalent of the tropopause, a boundary below which temperature ceases to decrease as depth increases. Like the tropopause, the thermocline severely restricts mixing between water above and below it. Once carbon-based matter - or anything else - sinks below the thermocline it may be hundreds or even thousands of years before it crosses back into the upper region from which it has a chance of being returned to the atmosphere.

The absorption and release occur in different places. This is partly because the amount of carbon dioxide that can dissolve in water increases as the temperature of the water falls. The gas dissolves in high latitudes and is released in equatorial regions. It is absorbed more in winter than in summer and is released more in summer than in winter for the same reason. The regional disparity is also due to the effect I mentioned earlier, the increased salinity of water immediately beneath sea ice that makes it sink and so increases the rate of mixing between surface and deep waters.

Considerations such as these lead to concern about what will happen in future. If sea ice melts over large areas, reducing the mixing effect, and if the water near the sea surface becomes markedly warmer, the oceans will absorb less carbon dioxide. They could even 'flip' from being a net 'sink' into which carbon dioxide disappears to being a net source, releasing it. A point could be reached when the oceans themselves caused a major surge in the atmospheric carbon dioxide concentration.

Nor is it known how living organisms will respond to changes in temperature. We would like to imagine that as their environment grows warmer and as the amount of carbon dioxide increases they will grow and multiply more rapidly. Unfortunately, it is not necessarily so. Shortage of carbon dioxide is not usually a problem for marine organisms. Their growth is more often limited by shortages of mineral nutrients, and cold waters - upwelling from great depth and bringing mineral nutrients with them - are more biologically productive than warm surface waters. Oxygen, too, dissolves more readily in cold than in warm water, and is needed for respiration.

At present it is predicted that, unless steps are taken to

reduce the rate of increase, by 2030 – forty years from now – the atmospheric carbon dioxide concentration will have increased from its present 350 parts per million to between 400 and 550 parts per million and within less than a century it will double, to about 700 parts per million. If it doubles and if, as is also predicted, the consequence is that temperatures in the polar regions rise by 5 to 8°C (9 to 14°F), the winter sea ice would disappear and the rate of accumulation would accelerate.

If carbon dioxide is the most important greenhouse gas, methane is the second in order of importance. As I have mentioned, it may be implicated directly in the chemistry of the stratosphere quite apart from its role in the troposphere.

Methane

Methane is very much a biological product. It is possible to manufacture it from carbon dioxide and water, the obvious raw materials, but the process must go through a series of steps and at each step the intermediate product is very unstable. Bacteria produce methane inside their cells, where reactions are protected from disturbance, and the bacteria themselves live in places where there is no free oxygen because that is poisonous to them. Sometimes methane can accumulate, but it is always associated with things that are or were living.

In coal mines methane can be released by the mining operation and when it mixes with air it forms fire-damp, a very explosive mixture that has caused many mining disasters. Associated with petroleum, we call it natural gas and burn it as a fuel. It forms in the airless mud on the bottom of stagnant ponds and marshes and sometimes bubbles to the surface as marsh gas, which can ignite to burn with a ghostly blue flame and frighten the superstitious on dark nights.

These are all fairly minor sources. We burn methane in large amounts, of course, but burning destroys it – although the combustion releases carbon dioxide and water vapour – so releases of methane itself occur only through the leaks that are inevitable when it is mined, piped and transported. Rubbish dumps form another minor source which has received some publicity over the last year or two. We dispose of our household refuse mainly by burying it in landfill or controlled tips. The wastes are dumped, usually into a hollow in the landscape, and at intervals they are covered with earth so eventually all traces of them disappear. Beneath the surface those that can decompose do so, but the bacteria which do the decomposing include some that produce methane. The methane is sealed below ground where it collects until there is so much of it the gas begins to leak. When this happens it constitutes a very real fire or explosion hazard; waste dumps have burst into flames, apparently spontaneously, and they have been known to explode. Those that do not explode or burn nevertheless release methane into the air. Once it is well mixed with the air it is quickly diluted to concentrations that cannot ignite, but it still can and does absorb radiation.

At any rate until recently, the biggest source, oddly enough, was termites. Termites eat wood, but they cannot digest it unaided any more than you or I can. To help them they have colonies of microbes in their guts and these microbes include some that release methane. If there is more dead wood available and the world population of termites increases, then larger methane releases must surely follow.

The main source of the increase in the amount of methane in the air, however, is farming, in particular the growing of rice in paddies and the husbandry of ruminant animals, mainly cattle.

Wet rice-growing is a very efficient way to produce food and it has increased dramatically over the last 20 years or so as the so-called green revolution has made it possible for Asian farmers to grow two or even three crops a year, with a heavier yield from each, on land that formerly grew only one. The total world yield of paddy rice increased by more than 25 per cent between the period from 1974 to 1976, (which is averaged to provide a base figure) and 1983, on an area of more than 1.5 billion square km (600,000 square miles).

In paddy-rice production artificial ponds are flooded and rice seedlings sown in the mud on the bottom. Then, when it is time to ripen the crop, sluices are opened and the water drained away from the paddies, which become small fields. In the mud there are methane-producing bacteria and the methane is released when the paddies are drained. The rice plants themselves also release some methane. As paddy rice production has increased, so has the amount of methane being released.

Elsewhere in the world, rising prosperity has increased the demand for meat, and the number of cattle has grown to meet that demand. Today there are about 1.5 billion head of cattle in the world. Like termites, ruminant mammals have microbes in their guts to help them digest the tough celluloses that make up the cell walls of grasses, and some of those microbes release methane which the animals excrete at up to 250 grammes (half a pound) of methane a day each.

The amount of methane in the atmosphere is increasing at about 1 per cent a year, so it could double in about 70 years. Over the last century the concentration has risen from about 900 parts per billion to about 1,700 parts per billion and by 2030 there could be 2,200 to 2,500 parts per billion.

The release of methane could also surge suddenly. Over vast areas of northern Canada, Alaska, Europe and Asia the ground is permanently frozen. In summer the surface thaws and plants grow, but a little further down the ice never melts. This is permafrost and it is frozen because the soil is waterlogged. Should the temperature rise in these high latitudes the permafrost might begin to thaw. Should that happen the permafrost areas will turn into swamps and marshes, their bacterial populations will grow more active, and marsh gas - methane - will be released. The quantity may not be large for each acre of land but the number of acres by which that small quantity must be multiplied is so large that the total methane output might well exceed all other sources.

CFCs, Ozone and Nitrous Oxide

CFCs are the third most important group of greenhouse gases, ozone is the fourth, and nitrous oxide (N_2O) is the fifth. About 19 million tonnes of nitrous oxide enter the air every year from natural sources, human activities add some six million tonnes more and, as with the other greenhouse gases, the concentration is increasing. A century ago the air contained about 285 parts per billion of nitrous oxide, today it contains about 310 parts per billion and by 2030 this is expected to have risen to 330 to 350 parts per billion.

All proteins contain nitrogen compounds, so when material of plant or animal origin is burned oxides of nitrogen, of which nitrous oxide is one, are released. The burning of forest vegetation, grassland, and also the burning of stubble in British fields, all produce nitrous oxide. It also enters the air as a result of using nitrogen-based fertilizers.

Commercial fertilizers supply plants mainly with three essential nutrient elements, nitrogen, phosphorus, and potassium. Phosphorus and potassium compounds are not very soluble in water, are taken up slowly by plants, and so only relatively small amounts of them are required. Nitrogen, on the other hand, is taken up by plants in the form of nitrate, and nitrate is highly soluble in water. Because of its solubility losses are inevitable because some of the fertilizer is simply washed away as water moves through the soil. For this reason nitrogen applications are high although there is not much difference in the amounts of nitrogen and potassium plants need. It is some of the surplus nitrate (NO_3) that is converted to nitrous oxide (N_2O).

There is no doubt that the atmospheric concentrations of carbon dioxide, methane, ozone, and nitrous oxide are increasing and that the concentration of CFCs will continue to increase until the steps that have been taken to restrict its production and use become effective. Nor is there any doubt that these substances intercept long wave radiation and that such interception could lead to a warming of the lower part of the atmosphere.

It is one thing to calculate an effect on air, however, as though it were sealed in a bottle under strict control, and quite another thing to calculate an effect on the global climate. It does not necessarily follow that a system as complicated as the climate will behave in the way the predictions suggest it should behave, and if a major change does occur it will be far from easy to detect it in its early stages. It is time, then, to try and see whether or not a greenhouse warming has actually begun.

CHAPTER 6

Living in a Greenhouse

In 1986 my family and I went on a camping holiday to Dyfed, in the far south-west of Wales. One advantage of a camping holiday is the absence of news from the world outside. The weather was not very good that year but, lacking news, we were quite unprepared when it took a dramatic turn for the worse. For the whole of one day sheets of rain, lashed by a gale, immediately drenched anyone foolhardy enough to step outside the tent. Then, in the early evening, the rain stopped, the sky cleared, and the weather was perfect. The following morning we were woken early. One corner of the tent was flapping, its peg ripped from the ground by a howling wind. There was not much rain, but the gale screamed and plucked and tore all day. My most abiding memory is of the ceaseless, deafening noise of the wind. Then, once again, it died in the early evening and this time when we went outside into the quiet calm we could see the circular pattern of the clouds. We had sat it out through a dying hurricane, called Charlie, which had passed directly over us. The calm of the first evening was the eye of the storm.

The autumn of the following year, 1987, brought another hurricane, this time to the south of England where it caused great devastation and received much more publicity. Britain, everyone had thought until then, is not in a hurricane area. Hurricanes do not happen in our sleepy countryside. If hurricanes could strike us two years in succession, clearly something was very wrong with the weather.

Hurricanes Gilbert, in 1988, and Hugo, in September, 1989, came nowhere near Britain, but they were among the most severe hurricanes ever recorded and as Hugo swept through the Caribbean, Richard Anthes, a meteorologist at the National Center for Atmospheric Research in Boulder, Colorado, said that rising sea temperatures, leading to increased evaporation, could be expected to increase the frequency of hurricanes and that hurricane wind speeds might increase by as much as 25 per cent. Dr Kerry A. Emmanuel, of the Massachusetts Institute of Technology, has also studied the relationship between tropical storms and sea-surface temperatures and found them to be tightly linked. He has said that the temperature rise following a doubling of atmospheric carbon dioxide might increase the intensity of hurricanes by 40 to 50 per cent. This, then, is one of the predicted consequences of a greenhouse – or any other – warming so it is hardly surprising that these alarming hurricanes are seen by some people (though not by Dr Anthes or most other scientists) as evidence that the greenhouse effect has begun.

The Effects of Warming

Sometimes an overall warming may produce effects that seem paradoxical, and perhaps Britain has not escaped. In January, 1987, for about two weeks, Britain along with the rest of western Europe experienced a spell of intensely cold weather and in Britain about two thousand people died as a direct consequence because their homes were inadequately insulated and they were too poor to heat them sufficiently. The weather was called Siberian, and for once the popular name was accurate. A little bit of Siberia had visited us.

Eurasia is by far the largest of the world's land masses. In winter, when it cools, its central regions become very cold indeed. The cold pole is the place in each hemisphere where temperatures are lowest. The South Pole is colder than the North Pole, and there are places in Antarctica where winter temperatures of about −90°C (−130°F) have been recorded. The cold pole in the northern hemisphere is at a place called Verkhoyansk, in north-eastern Siberia, where the average temperature in January is around −50°C (−58°F) and the January temperature can fall to −68°C (−90°F).

This extremely cold air is very dense and it forms a large area of high pressure. Warmer air, passing over the Atlantic and then moving eastward across the continent, encounters the cold, dense air and rides up over it. This tends to push against the cold air and prevents it from spilling westward, the confining force being proportional to the difference in temperature between the warm and cold air masses. In most winters, however, the system relaxes briefly now and then and allows a tongue of this Siberian air to extend into Europe as a ridge of high pressure, but that is not what happened in 1987. That winter the Siberian air was just a little warmer than usual - warmer by Siberian standards, of course, and not by ours. This reduced the temperature difference, the gradient, between the warm air to the west and the cold air to the east, and the warm air did not hold back the cold air quite so firmly. The whole of the Siberian air mass moved westward, pushing the warmer, western air in front of it. Siberia itself enjoyed unseasonably warm weather, from the Pacific, and so did Iceland as the Atlantic air moved back into the Atlantic, but western Europe received the Siberian air.

Strange though it seems, therefore, the intense cold from which Europeans suffered in fact was caused by a warming of the air in Siberia. It was a temperature rise, but we felt it as a fall. The question is: why was Siberia warmer? Some scientists, at the Climatic Research Unit at the University of East Anglia, think the warming may have been due to the greenhouse effect. At least, this kind of phenomenon is consistent with a general greenhouse warming.

We all know that, quite apart from hurricanes, in recent years the British weather has been more than usually dismal. For one thing, there were several years in succession - 1986, 1987, and 1988 - when we had no real summer. There was the occasional fine day, but most of the time it was cloudy, or actually raining, and distinctly cool. Was there a reason for our summerless years? There may have been. When the surface waters of the Atlantic are warmer than usual north-western Europe tends to experience cool, wet summers.

In a good summer - and an ordinarily cold winter - the movement of air across the North Atlantic tends to break up into a pattern of cells of high and low pressure and one of these high-pressure cells can become established a little to the west of Europe and then remain there, sometimes for weeks on end, with a ridge extending to the east. As areas of low pressure move eastward they are deflected to the north or south by this blocking high, and the ridge dominates our weather, making it typically anticyclonic - dry, and warm in summer or cold in winter. In the bad years, associated with higher sea-surface temperatures in the Atlantic but in ways that are not fully understood, the blocking high fails to become established and so the low-pressure depressions go their miserable way unimpeded. Cool summers, then, may be linked to a greenhouse warming of the ocean.

Good summers, on the other hand, when Britain, or parts of it, can go without rain for weeks or even a few months on end and the fields parch and reservoirs run dry, are simply part of the usual weather pattern. Between May 1975 and August 1976, for example, the south of England had only about half its normal rainfall, bringing the most severe drought since 1727, which is when records began to be kept. Again in 1989, when the run of poor summers came to an end, Britain suffered a drought. In both cases a blocking high had established itself and refused to budge. Under such conditions the ground dries so much there is little evaporation from the surface so fewer clouds form over land and the drought tends to perpetuate itself.

Southern Europe has also experienced unusual weather. Exceptionally dry weather led, in August, 1989, to huge forest, scrub and grass fires in the Mediterranean region. People were killed or injured in these fires in Sardinia and Corsica and buildings were destroyed and people evacuated around Marseilles, Draguignan, and Aix-en-Provence in southern France. Two hundred fires broke out on one day alone - 28 August.

Droughts

Then there are the real droughts, those that affect parts of the world where dry periods are measured in years rather than weeks and the cause is different. If the average temperature rises in high latitudes while remaining fairly constant in low latitudes, the polar air masses may be held closer to the poles. In effect, the belt of warmer, low-latitude weather, and everything associated with it, will expand. This might well cause deserts to move from their present locations, roughly in a belt around the world in both hemispheres, to the regions bordering them.

Does this explain the drought that afflicted Ethiopia some years ago, and the droughts in the Sahel region, along the southern edge of the Sahara? Again, some scientists think it may. The African drought may be linked to relatively high sea-surface temperatures in the southern oceans and in the northern part of the Indian Ocean, associated in turn with changes in the flow of the prevailing westerly winds. In other words, the drought appears to be part of a widespread rise in temperature, so it is possible that it is either evidence that the greenhouse warming has begun or at least is consistent with it.

It is not only Africa that has suffered from drought during the 1980s. The 1987 Indian harvest failed for lack of rain and in 1988, while in western Europe snuffling holiday makers in plastic raincoats wandered damply along sea fronts, the central United States and southern Canada baked and dried in what was said to be the worst drought for half a century. Crops withered, boats were marooned as water levels fell in what had been navigable rivers, and cattle were slaughtered because there was no food for them. The drought continued into 1989 and in May the Iowa National Guard was transporting water to farms. Meteorologists in Iowa said early in the summer that so far 1989 was the driest year on record. In March, dust storms in Kansas closed inter-state highways. The 1989 winter wheat crop in Kansas was predicted to be only 60 per cent of its normal size. The soil was so dry and the water table so low that such rain as fell simply soaked into the ground and vanished. Very naturally, people recalled the dust-bowl of the 1930s and they were afraid.

The American drought was serious, for American farmers obviously, but also for the security of the world. Farmers received government help and shortages raised prices until in 1989 they were almost double those paid a year earlier, and this helped the farmers. Meanwhile, however, the world stocks of food fell after the Indian harvest failure. Then the Argentinian crop was harvested and was 15 per cent lower than usual and the Soviet harvest was also very poor. The North American drought has to be seen in the context of this general decline in stocks. Unless these are replenished the overall effect will be to raise food prices everywhere over the next few years.

Until the failure of the Indian harvest in 1987 the world as a whole had reserves of food sufficient to last for 100 days. The stocks had never been higher. When harvests are poor, stored food is released for sale to hold down prices and when harvests are bountiful food is taken into store to support prices. Without this cushioning world food prices would fluctuate wildly and might spiral viciously. A good harvest would depress prices paid to farmers who then would be less able to afford seed and fertilizer for the following year, so yields would fall, thus raising prices again but in a chaotic way that would be gravely destabilizing - a factor worth bearing in mind if you are bothered about the EEC food mountains which in fact are used in the same way and are sufficient to last for only a few weeks. Without stabilization, when prices rose the poor would go hungry and the rich countries would experience increased inflation, damaging their economies.

Was the North American drought evidence of the greenhouse effect? At hearings into the drought in June, 1988, Dr James Hansen, a climatologist who is head of the Institute of Space Studies at NASA, told the Senate Energy and Natural Resources Committee he was almost certain of such a link and that the time had come to take positive steps to control a deteriorating situation. His view was widely publicized but other scientists were rather more cautious, although most of them agreed that the 1988-9 drought was typical of the kind of effect a greenhouse warming might be expected to produce. It was some months and a great deal of investigation and calculation later that the scientific view changed. The American drought was caused not by a greenhouse warming but by El Niño, the curious phenomenon in the Pacific that I have mentioned before and shall describe in the next chapter.

Freak Weather and the 'Signal-to-Noise' Ratio

While Britain was cool and damp and North America parched, other parts of the world were drenched. In September, 1988, there were unusually heavy rains in Australia, and severe flooding in southern China, Bangladesh, and Sudan. This weather, too, was blamed by some people on the greenhouse effect and by others on La Niña, sister to El Niño.

One by one such examples of unusual or even freak weather add up and they all seem to point in the same direction. The proper conclusion appears obvious. We may not be able to prove it yet, but it is entirely reasonable to suppose, with Dr Hansen, that the greenhouse warming has begun. Once we accept this, other pieces of evidence start to accumulate. Almost casually the warm British summer of 1989 came to be called a greenhouse summer. It is less likely that yet another cool, wet summer would have earned such a name, but the link with a warm Atlantic might have justified it.

Perhaps the greenhouse warming has begun, but there are grave dangers in accepting such evidence as this. In the first place, our memories are highly selective. Most of us remember the long summers of our childhood. It is natural that we should. They were the summers we most enjoyed, literally the memorable ones. We forget the poor summers when we had to stay indoors watching the rain running down the windows and desperate with boredom. We remember the really hard winters for the same reason.

An information scientist would explain this in terms of the signal-to-noise ratio. A sight or sound constitutes information only if it tells us something we did not already know. Switch on a crackly old radio and the crackles are background noise (sometimes called white noise), conveying no information, and you have to twiddle knobs or strain your ears to hear the broadcast which makes sense and so constitutes the signal – the part of the total sound that conveys information. Unless the signal is louder than the noise – which is where the ratio comes in – you will hear nothing useful at all. It can be more subtle than this. If someone tells me my name is Michael Allaby, this is not information, just noise, because I know my name already. Similarly, ordinary, everyday weather conveys no information to us. It is only unusual weather that arouses our interest, that supplies information, and so it is only unusual weather that we notice and remember. The unusual weather is the signal, the ordinary, everyday weather mere noise.

The trouble is that the signal can mislead because our memories are limited. We cannot remember things that happened before we were born, for example, and even during our lives our memories are neither perfect nor permanent. So what constitutes unusual weather? The period from June to September, 1933, was the second warmest (the other was 1911) since before 1881, with the highest temperature, of 34°C (94°F) at Cambridge on July 27, and in much of England and Wales December was the coldest since 1890. Some people thought 1911 was the hottest and driest year ever, with a temperature of 38°C (100°F) recorded at Greenwich on August 9. In 1762, a snow storm lasted for 18 days in February and in some places the snow lay up to 12 feet deep. That was during the Little Ice Age, of course, but 1762 was also a year of intense heat and severe drought. The fact is that in Britain, and probably in most countries, some weather record is broken almost every year and now and then there are years of extremes of several kinds. 1947 was such a year. The winter really began in late January, very cold - the sea froze in some places - and with heavy falls of snow, and lasted until well into March. Then there was heavy rain which, as the snow melted, caused much flooding. In the summer there were very high temperatures and drought. February was the coldest month in England and Wales since 1895, March one of the wettest ever recorded, and August the warmest since before 1881.

It begins to look as though our obsession with the weather makes every year a freak and really it is the normal years that are uncommon. This is obviously a contradiction in terms but it illustrates the difficulty of reading any significance into the weather of a single year, or even of a short sequence of years.

When we detect what we think is freak or record-breaking weather we begin at once to search for explanations, and the greenhouse effect has competitors. If the 1988-9 North American drought resembled the dust-bowl drought of the 1930s, what caused the dust bowl? Environmentalists used to believe it was bad farming leading to severe erosion but scientists dismissed that idea long ago. The farming was not all that bad, and the erosion was caused by drought. The drought was the real problem.

One explanation at the time invoked radio waves, though it is a little difficult to see how they could have produced any effect. There was another drought in North America in the 1950s. That time the atmospheric testing of nuclear weapons was blamed. Theoretically it is not impossible that weapons testing might have produced a climatic effect had the tests injected really large amounts of particles into the stratosphere, so the explanation is not absurd. There is no reason to suppose, however, that in reality tests produced anything like the quantity of stratospheric particles needed and, unfortunately for the explanation, if they had it is unlikely that they would cause drought. By increasing the albedo of the planet they would exert a cooling effect, producing cool, wet summers. Indeed, Professor P. Handler, a physicist at the University of Illinois, links the 1988-9 drought to stratospheric particles, or rather the lack of them, but from volcanoes, not weapons tests.

In terms of human casualties, one of the worst volcanic eruptions in history occurred on November 13, 1985, in Colombia. In the previous December the 5,400 metre (17,700 foot) volcano Nevado del Ruiz began to emit gases and there were tremors within the mountain in the months that followed, with a small explosion in September before the major ones in November. The big eruption comprised two explosions. They melted part of the ice cap and sent two streams of melt water mixed with rock and mud - technically lahars - hurtling into a valley, bursing a natural dam and causing the deaths of more than 25,000 people. The eruption also ejected large amounts of particulate matter, and there may also have been a small contribution from Mount Etna, which erupted on Christmas Day of that year. According to Professor Handler, the stratospheric particles from Nevado del Ruiz caused a slight climatic cooling and the US produced excellent corn (maize) crops in 1986 and 1987. By 1988, he said, the particles would have been removed, the stratosphere would have been clear, and the weather would have been hot and dry.

It sounds convincing, but the Mexican volcano El Chichón, whose eruption in 1982 injected very large amounts of particulate matter into the stratosphere, seemed to produce no agricultural benefit in 1983 - a heat wave and drought lasted from June to September over much of the United States, and was attributed to our friend El Niño. This does not quite dispose of Professor Handler's theory, however. El Chichón did produce a very slight cooling and some scientists believe it would have been much more pronounced had the climate not been warming at the time. There is also the opposing view that in general a warming brings wetter conditions and drought is more typical of a cooling.

Other scientists link the weather to changes in solar output and the sunspot cycle. According to them, the weather generally should start to become drier from the autumn of 1987, reach its driest in 1992, and then be rather wetter for the following five or six years.

Do we need to look for explanations at all? Humans have been keeping records of the weather for a century or two but some trees have been keeping them for very much longer. Bald (or swamp) cypresses (*Taxodium distichum*) grow naturally in North Carolina and studies of their annual growth rings have provided climatic data, year by year, back to AD 372. The tree rings show that throughout this long period the local climate has changed over a time scale varying from 21 to 63 years and averaging 30 years. In other words, the climate changes according to a fairly regular cycle. Over the centuries the studies have detected 28 occasions when for some years the climate was markedly wetter or drier than the long-term average. One of these wet periods began in 1956 and may have been coming to an end in 1985 and 1986, giving way to drier weather. The 1930s dust bowl occurred during a previous dry episode.

Melting Ice Caps

If the climate really is warming all over the world, sooner or later the polar ice caps will start to melt. That, you might be forgiven for thinking, should be easy enough to measure, and the polar ice is being watched very carefully.

When large bodies of ice start melting two things are likely to happen. The first is that at its weakest points the sea ice breaks, releasing (the technical term is calving) icebergs. The second is that the sea ice becomes generally thinner.

Once again there have been some spectacular events. In October, 1987, for example, an iceberg broke away from the Ross ice shelf on the coast of Antarctica. Icebergs often break away from the edges of ice sheets in spring (October is the southern-hemisphere seasonal equivalent to April in the northern hemisphere). This iceberg, however, was a giant, 174 kilometres (108 miles) long, 40 kilometres (25 miles) wide – three-quarters the size of Cornwall – and on average about 230 metres (750 feet) thick. It was so large that scientists estimated it would take ten years to melt.

The West Antarctic ice shelf, of which the Ross ice shelf – itself about the size of France – forms part, is believed to be unstable. The area consists mainly of a low-lying archipelago whose islands lie beneath and are linked together by very thick ice. The ice tends to break up a little in summer, calving icebergs that drift away to sea, and thicken in winter, and the prevailing winds drive the ice as it forms causing it to pile up in particular places, especially in the western part of the Weddell Sea. Icebergs are common and startling, and though the dimensions of the 1987 Ross sound huge, it is not the largest iceberg known. Over the years others have been measured that were more than 200 kilometres (124 miles) long and rose up to 70 metres (230 feet) above the sea surface.

The instability of the shelf makes it the region of the Antarctic that is most closely observed. A quite small change in temperature, wind or ocean current might cause it to start breaking up and should that happen the amount of fresh water that eventually would be released as the icebergs melted would raise sea levels everywhere by 4 to 5.5 metres (12 to 18 feet). If all the ice in Antarctica were to melt, world-wide sea levels might rise by as much as 37 metres (120 feet). Scientists do not believe this will happen and it is possible that a global warming could make the ice sheets thicker, so removing water from the oceans, because there would be more precipitation, falling in the Antarctic desert as snow that would accumulate.

The presence of icebergs, even huge ones, means nothing. It certainly does not mean the ice sheet is breaking up or melting, but there is other evidence that may be more disturbing.

In August, 1988, two new items of information were given to scientists attending a meeting of the Geographical Congress held in Hobart, Tasmania. The first, from Dr Ronald Smith, head of the terrestrial ecology unit of the British Antarctic Survey, was that at least some of the glaciers were retreating. The second, from Dr Ian Allison, of the Australian Government's Antarctic Division, referred to satellite photographs of the continent showing a retreat of the edges of the ice cap. He also said that records collected over 30 years showed a rise in temperature along the coasts of Antarctica.

A year later, in September 1989, further evidence was released. It had been proposed to build an airstrip near the British station at Rothera, on the west coast of the Antarctic Peninsula, and an assessment had to be made of the environmental consequences of doing so. The report showed that ice was being lost rapidly near Rothera and the average summer temperature had risen about 1°C during the 1980s and the rise seemed to be accelerating, although so far the year-round temperature had changed little.

There seems to be general agreement that the Antarctic ice is retreating and has been retreating at least since the 1950s, but this may have nothing to do with any greenhouse warming. Antarctic temperatures rise and fall according to a natural cycle and in the late 1980s they reached their maximum. It may be that the greenhouse effect has influenced the rise but it is far from certain. Temperatures have risen much higher in the past. The traces and remains of vegetation on ground that is ice-free show that two centuries ago there was much more vegetation and much less ice than there is today, so the climate must have been markedly warmer. The relatively high temperatures at present are nothing out of the ordinary. Over the next few years the natural cycle should bring lower temperatures. If they do not fall, or if they actually rise, this may be firm evidence of a greenhouse warming.

Meanwhile, at the other end of the world, more observations were being collected. North of Greenland, where the sea is covered permanently by ice, the ice was growing thinner. In 1976 its thickness varied from six to seven metres (20 to 23 feet) but in 1987 it was only four to five metres (13 to 16 feet) thick.

The evidence for warming is convincing but the interpretation of the evidence is difficult because even now so little is really known about what happens in polar regions. More is being discovered almost daily and sometimes it complicates or even contradicts ideas that once seemed beyond challenge.

The albedo effect of the ice caps is obvious and no one questions it. Should the ice caps melt, the white ice and snow would disappear to be replaced by water in the Arctic and water and rock in the Antarctic, and the consequent reduction in the planetary albedo would mean more energy was absorbed by the surface and less reflected, so favouring further warming. It follows, therefore, that on the small scale the same thing should happen, but to a lesser degree: the melting of a little polar ice must lead to a little warming. It turns out to be not quite as simple as that. The system resists change.

There are breaks in sea ice, channels up to 500 metres (550 yards) wide that open to expose liquid water, then close again or are rejoined as the sea freezes. The breaks are called leads and they come and go unpredictably. They are caused by stresses in the ice produced by the wind and they were difficult to count or measure until, quite recently, new instruments were introduced that allow overflying aircraft to record their coming and going.

When a lead opens a great plume of water vapour rises into the air, sometimes to a height of four kilometres (13,000 feet). Polar air, like air anywhere else, contains nuclei on to which water can freeze and so ice crystals form. The change from vapour to solid releases the latent heat of sublimation, warming the surrounding air which then rises, keeping the ice aloft, and ice clouds form. These clouds are very difficult to see from above because they are white against a mainly white background and, of course, they are invisible during the long night of the polar winter. As spring arrives they provide a high-albedo layer that has no effect at all unless there is a break in the ice below them and the sea is revealed. When that happens the albedo remains high, because of the clouds, rather than falling, and so the warming effect is reduced. This may mean that the observed reduction in ice cover marks a bigger temperature rise than had been supposed. If a general warming is taking place, therefore, because of the greenhouse effect or for any other reason, the ice caps are perhaps delaying it.

Outside the polar regions glaciers are confined to mountainous areas high enough to have low temperatures all year round. Mountain ranges occur in all latitudes, and so do glaciers, and throughout the world glaciers have been retreating since the 1830s. This may reflect nothing more than the gradual ending of the Little Ice Age, a climatic episode from which we may still be emerging. The glacial retreat does coincide, however, with the increased release of carbon dioxide that began soon after the factories of the industrial revolution converted from water-power to coal-generated steam power, so it could be a greenhouse effect. Or could it be that the ending of the Little Ice Age was itself brought about by increased atmospheric carbon dioxide, that the greenhouse warming actually began long ago but was not attributed to this cause? We may ask questions, but no one is able to answer them.

Overall Trends

We can sum up the evidence by saying that the average temperature over the whole world began rising during the last century and the rise is well recorded from about 1920 until about 1940. Then there was a cooling that lasted until about 1970, most strongly marked in the Arctic and associated with a steepening of the thermal gradient between the equator and the North Pole. The same change probably occurred in the southern hemisphere but the records are less complete because there is less land in the southern hemisphere and so reports from ships are more important than they are in the northern hemisphere - and there are fewer ships. Since then the rise has been resumed, but with two cooler years, 1984 and 1985, which followed and may be attributed to the 1982 El Chichón eruption.

Over the last century the average world-wide temperature has risen by about 0.5 °C, and the warming has affected both hemispheres more or less equally and has been accompanied by changes in pressure and rainfall patterns. Over the last 30 to 40 years precipitation has been increasing in the middle latitudes and decreasing in low latitudes. In Europe the increase began in the last century and it is even more marked in the Soviet Union. In North America there was a decrease in precipitation from about 1880 to the 1930s but since then it has been increasing there, too. In New South Wales, the spring, summer, and autumn rainfall in the period between 1946 and 1979 was 30 to 40 per cent higher than in the period from 1913 to 1945.

That such a small average rise took a century is a little misleading because during the 30 year cooling the temperature fell back a long way. So the fact that it has recovered in less than 20 years suggests it may be accelerating. In March, 1989, H. Flohn and A. Kapala of the Bonn Meteorological Institute – two of the world's most eminent climatologists – reported that the average temperature in the troposphere over the tropics had risen by nearly 1 °C since 1965 and over the equatorial Pacific the amount of water vapour had increased between 20 and 30 per cent.

This much we know from measurements and records and we can add a little speculation. Why was the warming halted between 1940 and 1970? That was the period during which the factories of the industrial countries of the northern hemisphere were engaged first in supplying weapons and materials for the war, and then in the task of post-war reconstruction. It was a time of furious industrial activity and the industry was a great deal dirtier than industry would dream of being today, partly from necessity but also because the dangers of atmospheric pollution were less well known. The burning of fuel released carbon dioxide, certainly, but it also released sulphur dioxide in very large amounts, along with dust and other fine particulate matter. These emissions supplied condensation nuclei and perhaps led to an increase in cloudiness, so cooling the surface. If this, or something like it, provides a plausible explanation for what happened, then perhaps we should conclude that the halting of the warming was due to human intervention and otherwise would not have occurred.

In Britain this general rise in temperature has also been observed indirectly because it produces other effects. Aphids, for example, are insects about which a good deal is known because they are of great economic importance as pests. They can reproduce either sexually or asexually. Usually, females lay fertilized eggs in autumn on particular plants. In spring, as the weather grows warmer and suitable food plants start growing, the eggs hatch to produce females that move to nearby food plants. These females produce young without mating – parthenogenetically. The young are born alive and active and all of them are wingless females that also produce female offspring parthenogenetically and the offspring, too, can reproduce in the same way. Producing young like this at the rate of a few a day each, generations succeed one another very rapidly and the aphid population explodes – which is why aphids are such serious pests. In late summer winged forms are produced, some of which are male, mating occurs, eggs are laid, and all the aphids die.

In exceptionally warm years, however, the adult aphids survive and continue breeding asexually throughout the winter, which means there are more of them at the start of the next crop-growing season. The infestation is more severe, and so people notice it. In recent years over-wintering adult aphids have been recorded more often and further to the north than they used to be.

The 1980s have been the warmest decade for a hundred years and 1980, 1983, 1987, and 1988 were among the warmest years on record. Figures for 1989 are not yet available but there seems no reason to suppose it will not continue the trend.

Nor is there any doubt that sea levels have been rising since about 1900 at the rate of 10 to 25 centimetres (4 to 10 inches) a century. The rise is recorded by tidal gauges and is important commercially because it affects access to ports. Sea level can change locally because of movements of the Earth's crust associated with isostatic readjustment following the end of the last ice age, and because of coastal erosion. The world-wide rise, however, is believed to be due mainly to the fact that the seas are growing warmer and the water is therefore expanding.

The final item of evidence is slender but it concerns measurements of the radiation balance of the Earth – the amount of radiation received from the Sun compared with the amount reflected or re-radiated back into space. In August, 1987, a Vancouver meeting of the International Union of Geodesy and Geophysics was told by Dr Wayne Evans, of the Atmospheric Environment Service of the Canadian Government agency Environment Canada, that he had made such measurements. Since 1975, he said, the amount of radiation being retained by the Earth had increased by 0.1 per cent, and he claimed this was the first direct evidence that the greenhouse effect had begun. His measurements covered only that part of the solar spectrum in which CFCs absorb, and so he attributed the increase to CFCs. If carbon dioxide, methane, or other greenhouse gases are interfering with the radiation balance the total figure will be higher, but so far no one has been able to measure it.

Has the greenhouse effect begun? At present it is quite impossible to say. Certainly the world is growing warmer, sea levels are rising, and the ice is retreating. These changes have been measured, but the value of the interpretations we can place on them is limited. They are consistent with a greenhouse warming but they fall far short of proving it and there is no shortage of alternative explanations for weather that strikes us as extreme, such as the North American and African droughts.

So far, all the changes that have been observed fall within the range of ordinary climatic phenomena for which we do not really need special explanations. The half-degree temperature rise, for example, is sufficient to explain the rise in sea level and the retreat of glaciers and ice shelves, but it is very small. The difference in average temperature between an ordinarily warm year and an ordinarily cool year is between 1 and 1.5°C - twice to three times the measured temperature rise - although, of course, the present rise appears to be sustained over a number of years rather than being an isolated, one-off event.

It will be another 10, or perhaps 20 years before anyone can be certain whether or not a greenhouse warming is taking place. If the greenhouse effect were of purely academic interest this might be frustrating to scientists but it would be of little importance or interest to the rest of us. Unfortunately it is not that kind of change. Its consequences may be profound and extend far beyond the realm of our daily weather. For some people in some parts of the world a warming might well prove beneficial but for others it might be very grave indeed and by the time we know, beyond any reasonable doubt, that the change is taking place it may be too late even to limit it, far less to reverse it.

For politicians and political campaigners it is all very unsatisfactory. It sounds as though we should take certain action now, and this is what campaigners will urge - but what action? How can we respond appropriately unless we know more, much more, about the situation we face, and whether we face a dramatic change in the climate at all?

It is time to look more closely at what seems to be the joker in the pack - El Niño and his sister La Niña. Then we can consider the way climatic predictions are made and what those predictions tell us may happen to our living conditions in years to come.

CHAPTER 7

What Will the Weather be like for our Grandchildren?

There is no disagreement among scientists about the chemical composition of the atmosphere. The presence of greenhouse gases is not in dispute, nor is the rate at which they are being released and are accumulating. What will happen to their rate of release in the future obviously depends on the way humans respond. What will happen to their rate of accumulation in the atmosphere will remain much less certain until we find answers to a number of important questions. Will the oceans continue to absorb carbon dioxide at the present rate? Will the rate of absorption change, or under certain conditions will the oceans begin to emit the gas? Will the melting of permafrost release methane, and if so, how much? Have forest clearance and the burning of vegetation caused the world's vegetation to become a net source of greenhouse gases, emitting more than it absorbs? If not, is this likely to happen and, if so, when?

The usual description of the greenhouse effect makes it sound rather simple: if you pump certain gases into the air the entire planet will warm. By now you may be starting to think that the subject is anything but simple, and you are right. Indeed, it is extremely complicated - perhaps complex as I suggested earlier - and all the predictions about it are uncertain and must be heavily qualified. This does not mean they are valueless or that we can afford to ignore them, only that we should see them in their proper context. It may be easier to achieve that if we can understand how they are produced and the way scientists use them.

Climatology and Meteorology

Here I should describe the branches of science that are involved. Clearly, predictions of what the climate will be like 50 or 100 years from now are not the same thing as weather forecasts. Weather forecasts, detailed but for only hours or days in advance, are made by meteorologists. Their scientific discipline, meteorology, is the study of the weather and how it is produced. The word is from the Greek *meteoros*, meaning 'lofty', and *logos*, meaning 'discourse'. The scientists who study climates, rather than the detail of the day to day weather, are climatologists. The two disciplines overlap, of course, because you cannot study climate unless you understand how weather patterns are formed and develop and you cannot understand those patterns unless you understand the principles underlying the study of climates. Meteorologists and climatologists often work together in the same institution. Greenhouse predictions are the work of climatologists rather than of meteorologists.

Science proceeds, at least in principle, by making what you could call best guesses in order to explain things. These are not ordinary guesses, plucked from the air and backed by nothing more substantial than intuition, but neither do they claim to be statements of absolute truth and they can be altered, or abandoned entirely, if someone produces a better guess. The process often begins with an observation of something in the real world. Then attempts are made to explain the observation. The explanation allows predictions to be made on the lines of 'A happens because of changes in B, so if you change B, A will happen'. This leads to experiments designed to test the explanation. If the anticipated result happens the explanation has won that much support. If it does not happen the explanation may have to be reviewed, modified, or abandoned. Or, instead of an observation, the process may begin with an idea.

In our case the idea, first proposed long ago, was that there might be a link between the radiation balance of a planet and the presence of carbon dioxide and certain other gases in its atmosphere. Experiments are difficult because you cannot use an entire planet as a laboratory, but in recent years an alternative has emerged. It has become possible to construct 'models' of a planet for experimental purposes.

GCMs

I place the word 'model' in quotation marks because these are not physical models in the ordinary sense, like models that are miniature representations of aircraft, ships, or houses. They are mathematical and they live inside computers. Their construction had to await the invention of modern computers because, although a computer can perform no calculation that a human could not perform with pencil and paper, the number of calculations involved is so large the task would take longer to complete than the events in the real world they sought to represent. There would be little point in building a model to predict what will happen in 10 years' time if building the model took 15 years.

In the case of what are called global circulation models (or GCMs for short) of the Earth's atmosphere, the task would take a great deal longer than 15 years, for the number of calculations involved is truly vast. It is so vast, in fact, as to have become practicable only with the introduction of computers that can store almost unlimited amounts of information and perform millions of separate calculations a second, and even these supercomputers are stretched to their limits. The computers are very large and very expensive and so there are not many GCMs in the world. Some were built originally for long-range weather forecasting but have since been adapted for their new job. There are GCMs in Britain, at the Meteorological Office at Bracknell, and at the University of East Anglia's Climatic Research Unit, for example, and in the USA at the Geophysical Fluid Dynamics Laboratory of Princeton University, the Goddard Institute for Space Studies, and at the National Center for Atmospheric Research.

Inside a GCM the world is represented by a three-dimensional grid pattern. The horizontal dimensions of each box in the grid are several hundred kilometres and it is several kilometres high, so every place on the surface of the Earth lies inside one box and beneath several more, directly above it at various altitudes. To keep the amount of computation within manageable limits the calculations and predictions refer only to the points where lines of the grid intersect.

The calculations are based on the known physical laws that govern the behaviour of gases, and the numbers on which they work represent such things as the chemical composition of the atmosphere and the energy received from the Sun. The information the computer delivers may describe only the temperature at each grid intersection, but on more advanced GCMs it will also include details of such factors as humidity, soil moisture, and wind speed. The model is tested by feeding in information from the past to see whether the climate it predicts resembles the climate that was recorded historically. Then particular details can be altered - the concentration of greenhouse gases might be increased, for example - and the model will predict the outcome. Described in this way the modelling does not sound too formidable, but quite apart from the time it takes to write and perfect the programs, on the fastest supercomputers in existence it takes several hours to run a model in order to produce a simulated world climate for just one year.

The models are severely limited, partly for lack of computing power and partly for lack of knowledge. It is the lack of computing power that restricts the calculations to points at the intersections of grid lines hundreds of kilometres apart. This amounts to more than the kind of restriction of scale that makes the difference between a large-scale and small-scale map. A large-scale map, for the use of aircraft pilots, say, can omit many details which are simply irrelevant to the purpose for which the map is intended, and a small-scale map, such as walkers might use, can include much more detail. The world climate is not like that. It is the small-scale detail that adds up to the largescale climate, but the models have to omit much of this detail.

Cloud formation and precipitation, for example, occur on a scale of kilometres or tens of kilometres, not hundreds of kilometres. To compensate for this the models use information they can calculate, such as humidity and temperature, to provide values for average cloudiness and when this average exceeds a certain value the models assume it will begin to rain or snow. Cloudiness affects atmospheric temperature, of course, depending on the type, thickness and height of the cloud, but so far the models are unable to calculate these characteristics and so they cannot say whether the clouds they predict will have a warming or a cooling effect.

Vegetation also has an important influence on climate. Plants transpire water, so adding water vapour to the atmosphere, and they alter the albedo of the surfaces on which they grow. Most GCMs either ignore vegetation or include an allowance for it based on very simplified values. There are many other details of this kind, local but very important, about which the models must make assumptions that could prove incorrect.

Nor are the GCMs very good at interpreting the relationships between the oceans and the atmosphere. The oceans act as a kind of 'thermal sponge'. They soak up heat because they warm more slowly than the land, so if the atmosphere is warming any effect will be delayed while the oceans warm. Estimates of how long this will take range from 20 to 500 years. The time it will take for the oceans to warm depends mainly on the way they distribute their heat through currents. Ocean currents convey warm water from the equator into high latitudes and return cool water to the equator, but as the oceans warm the pattern of currents may change. Any change in ocean currents may alter the thermal sponge characteristics of the oceans and it will certainly produce major climatic effects. The computational difficulty of including ocean currents in a GCM is so great that many models omit the currents altogether and those that include them do so in a very simplifed form.

The consequences of this limitation are considerable. In Britain, for example, the climate is influenced strongly by the warm North Atlantic Drift which washes our western shores. If the climate over the whole world grows warmer and the course of the North Atlantic Drift remains unchanged, then the water it carries will become warmer and the British climate will probably become a little warmer and a good deal wetter. If the North Atlantic Drift changes course, however, the situation might be very different. Deprived of our warming current, while much of the world grew warmer it is possible that Britain might actually grow colder, at least for a time. After all, the last time the North Atlantic Drift changed course before reaching us we experienced the Little Ice Age. It is not only Britain whose future climate is so uncertain. The ocean currents affect the climate in many parts of the world and until the GCMs can take full account of them no regional prediction can be regarded as reliable.

El Niño

This difficulty in incorporating the effect of ocean currents is being resolved to some extent as more is learned about the most subtle sea change of them all: El Ninõ, or to give it its modern full title the El Ninõ-Southern Oscillation event, or ENSO.

It occurs in the Pacific Ocean, just south of the equator, and its effects are felt first down the western coast of South America, usually in December - mid-summer in the southern hemisphere, of course - which is how it earned its name: El Niño is Spanish for 'boy child', referring to Christ. It has an opposite which scientists have called 'La Niña', 'girl child'.

To either side of the equator the prevailing winds, the trade winds blow from the east - north-easterly in the northern hemisphere and south-easterly in the southern. The region where they meet is called the Intertropical Confluence, or ITC, and it is not precisely at the geographic equator. There is also a thermal equator, the line around the world where the temperature is highest. This moves a little with the seasons and the ITC tends to follow it but the ITC is also subject to more complicated movements so its strength and rather wavy pattern vary over periods of a few days.

The ITC is associated with a region of low atmospheric pressure, especially pronounced in the Pacific, where the ITC is also known as the Intertropical Convergence Zone (ITCZ). Driving the winds, in the subtropics of the south-east Pacific (along the western coast of South America) there is usually an area of high pressure and in the Indian Ocean, just to the west of Indonesia, an area of low pressure. Air moves through the lower atmosphere from the area of high pressure in the east (South America) to the area of low pressure in the west (Indonesia), then rises, returns in the opposite direction at high altitude, then descends to contribute to the high-pressure area.

This pattern of air circulation is mirrored in the ocean. The winds blowing across the surface set up a current flowing to the west and because the sea surface is warmed strongly by the Sun the current carries warm water. A current of cooler water, deep below the surface, flows eastward, carrying water back in the direction of South America. The effect is to pile up warm water in the region of Indonesia and produce a situation in which the depth of warm water is much greater there than it is off the South American coast.

Another current, the Humboldt or Peru current, flows northward close to the South American Pacific coast, bringing cold Antarctic water to the equator. The relatively thin layer of warm water off South America allows water from this current, some 5 or 6°C cooler than the surface water, to well up from below.

The cold water carries nutrients from the rich Antarctic waters together with others it has collected along the sea bed and, being cold, it is rich in dissolved oxygen. As it wells to the surface it provides food for a vast population of organisms, culminating, at the top of the food chain, with fish. At one time Peru had one of the world's largest commercial fisheries, based mainly on anchovies caught in the Humboldt current. Indeed, it is something of an understatement to call the Peruvian fishery large. The catch peaked in 1970 at more than 12 million tonnes of fish out of a world total of 70 million tonnes. The British catch in that year, just for comparison, was a little over one million tonnes.

It was not to last. In 1972 the Peruvian catch collapsed to less than five million tonnes and in 1973 it was rather more than two million tonnes. There was much speculation about what might have happened, most people (including me) favouring the view that the Peruvians had simply overfished the stock and the anchovy population had fallen dramatically as a result. We know now that is not at all what happened.

Every so often, for reasons that are not well understood, the ITC is displaced to the south, and the distribution of atmospheric pressure changes so the difference in pressure, and temperature, between the high and low pressure areas is reduced. This is the Southern Oscillation. When it happens the trade winds slacken and warm, surface water spills back from Indonesia towards South America, as though the entire ocean had been tilted. The depth of warm water increases off the South American coast and the upwelling, nutrient-rich, cold water no longer reaches the surface. The surface waters contain less food and oxygen so the populations of all organisms decreases and there are fewer fish. The failure of the Peruvian fishery was due to one of these tiltings of the Pacific Ocean.

It is this eastward spilling of warm water that South Americans call El Niño and its association with the Southern Oscillation leads climatologists to call it an ENSO event. An ENSO event affects more than fish. It brings northerly winds and heavy rains to coastal America, both North and South, drought to Australia, Africa, and the interior of North America, and widespread climatic disturbances elsewhere. The 1988–9 American drought is now clearly linked to an ENSO event.

ENSO years occur irregularly but some have been identified: 1925, 1941, 1957, 1965, 1972 and 1987 were ENSO years, and ENSO events now seem to be happening every two to seven years. It may seem improbable that the weather can be affected so strongly by events thousands of miles away, but these are not minor events. They involve a reversal in the flow of ocean currents and occupy the whole of the vast arena provided by the equatorial Pacific.

In other years the ordinary distribution of air pressure, winds, and currents is sustained but intensified. The winds blow more strongly, more warm water is pushed westward, and the result is La Niña, not associated with a Southern Oscillation, and bringing familiar weather but sometimes in a more extreme form. 1988 was a La Niña year and on the basis of it a warm, dry summer was predicted for Britain early in 1989.

As so often happens, the answer to one question merely raises another. If some of the extreme weather we have experienced in recent years was due not to a greenhouse warming but to ENSO events, what caused the ENSO events? Are they themselves linked to the greenhouse effect but in ways we do not yet understand? It is possible. It is also possible that the weather associated with ENSOs will become much more common. If the sea-surface temperature continues to rise, the effect in the tropics may be more pronounced in the cooler than in the warmer waters because if the warmer waters become still warmer they will simply evaporate more water, so shading themselves with clouds and tending to offset the warming. It is the cooler waters that will warm. Should this happen the temperature gradient across the equatorial Pacific will be reduced, trade wind speeds will fall, and warm water will cease to be pushed westwards so strongly. This will not be an El Niño but the consequences will be much the same and it could become the usual state of affairs.

Temperature Increases and their effects

Allowing for all these qualifications there is general scientific agreement that if the concentration of atmospheric carbon dioxide doubles, or there is an equivalent increase in the concentration of other greenhouse gases, the average, year-round surface temperature over the whole Earth will increase by somewhere between 3 and 5.5°C (5.5 to 10°F). The warming will not be distributed evenly. There may be little change at the equator, so the permanent ENSO may not materialize, but near the poles the temperature may rise by as much as 12°C (22°F). Such a doubling of the concentration of greenhouse gases might take a century.

On a shorter time-scale, the average temperature may have risen by about 1 or 2°C (2 to 3.5°F) by 2030 and by 1.5 to 2.5°C (2.7 to 4.5°F) by 2050. The temperature change sounds small, but a difference of only 5°C separates our climate today from that at the most extreme part of the last ice age, 18,000 years ago. Nor will this be the first time the Earth has been markedly warmer than it is today. There is nothing unique about the predicted temperatures themselves. The dangers are believed to arise from the rate of change rather than from the change itself – although even the rapid rate of change may not be without precedent.

Plants

The plants will respond, and are responding now, to the increased amount of carbon dioxide available to them. Carbon dioxide is a basic raw material for photosynthesis and despite our emissions it is still scarce. Horticulturists routinely release carbon dioxide inside glasshouses to increase plant growth, which it does. In the tundra of Alaska, plant photosynthesis is increasing for this reason and it is fair to assume that other plants will also benefit.

For a time, more vigorous plant growth may reduce the rate at which carbon dioxide accumulates in the atmosphere - although even the short-term effect may be offset by the release of carbon dioxide caused by the burning of vegetation - but it can provide no permanent sink for carbon dioxide absorption. There is a limit to the total mass of plants the Earth can support and when that limit is reached the absorption of carbon dioxide through photosynthesis will be balanced by the release of the gas through the decomposition of dead plant material.

Plants have pores, called stomata, in their leaves through which carbon dioxide enters the plant and from which water is released to evaporate - the process is called transpiration. In some plants the stomata close as the carbon dioxide concentration increases and there is evidence from plants stored in herbaria that modern plants have fewer open stomata than those which grew in the past. Closing the stomata under such circumstances does not reduce the rate of photosynthesis but it does reduce the rate of transpiration so, without any change in temperature, an increase in the atmospheric concentration of carbon dioxide may alter the rate at which water is transferred by plants from the ground to the air. Any climatic effect will depend on the type of vegetation - forests and grassland differ, for example, in the extent to which the air held among the plants mixes with air from outside - but broadly speaking less water may be taken from the ground, so it dries more slowly, while less is released into the air, so there is less cloud formation and precipitation.

There are two biochemical pathways for photosynthesis, called C3 and C4. Most plants are C3 plants but a few, including such important crop plants as maize and sugar cane, are C4 plants. C4 plants make much more efficient use of carbon dioxide and their evolutionary success is probably related to the decline in the atmospheric concentration of carbon dioxide over millions of years. These plants will not benefit from an increased availability of carbon dioxide, but C3 plants will, and under these circumstances they may thrive at the expense of C4 plants. The cultivated C4 plants may need extra protection from weed competition and care may be needed to make sure their wild relatives - needed in some cases as sources of new genetic material - do not become extinct.

Plants respond more slowly than animals to changes in temperature and in more variable ways. In Quebec, studies of plants growing in exposed, windy sites close to the limit for tree growth show they have responded much less than might be expected to the warming that has occurred since the end of the Little Ice Age.

This slow rate of response is of major concern to conservationists. There have been many major changes in climate in the past but by themselves such changes rarely lead to a mass extinction of species. Plants colonize new sites, die out in old sites, and over a time they migrate to more favourable climes. The onset of an ice age, for example, exterminates all vegetation beneath the advancing ice sheets, but species survive in lower latitudes and return when the ice retreats. As the last ice age ended the tundra vegetation of southern Britain moved north in the wake of the ice sheet, new species established themselves further south, from seeds carried by the wind or by birds, and the process of recolonization has been documented in some detail from pollen retrieved from layers of soil that can be dated.

It was a gradual process. Tree species migrate at something like 1 km a year. If the temperature increases at a rate many times faster than any previous warming, the fear is that plant species may not be able to migrate fast enough and many could be marooned in environments that have become hostile to them. It is unlikely that all species are able to move at the same speed and this could make the situation worse as the more adaptable species crowded out the less adaptable.

That is the fear, but there are two reasons for suspecting it may be unfounded. The first is that although different species may migrate at different rates, the historical evidence from past climate changes, obviously subject to the same differential rates of migration, shows that eventually stable systems become established with no serious loss of species.

The second reason is that studies of past climate changes suggest they have occurred very rapidly indeed and certainly as quickly as the change being predicted now. An ice age, for example, begins when the area of summer snow cover increases, and because of the increase in albedo it is then subject to a very strong positive feedback - the cooling increases snow cover which reflects more radiation and so accelerates the cooling. Cold periods also end rapidly, at least sometimes. The cold Dryas episodes that interrupted the warming at the end of the last ice age may have ended in a matter of a few decades. This rapid change caused no extinction.

The effect will be felt most strongly in the tundra regions, at present vast areas supporting a sparse vegetation of small, stunted trees and bushes and low-growing herbs. Bordered by coniferous forest to the south and sea covered by permanent ice to the north, the tundra is confined almost entirely to the northern hemisphere because - apart from Antarctica itself - the southern continents do not extend far enough to the south. The southermost tip of New Zealand is in about the latitude of Normandy, and Tierra del Fuego, in the far south of South America, is at about the latitude of Newcastle-upon-Tyne. As conditions suitable for coniferous forest move northward and the melting permafrost produces vast tracts of marsh, there will be nowhere for the tundra species to go. This does not mean there will be extinctions among tundra species, only that the total area of tundra will be reduced.

One conservation problem is real, however. There are plant communities, especially in the industrialized countries, that exist today only because they are protected as nature reserves or national parks, surrounded on all sides by land that is used or occupied by humans. They could not migrate of their own accord unless they were allowed to colonize adjacent land, which would be possible only if humans were prepared to stand aside and give them space.

Animals depend on the plant communities in which they live and although they can migrate rapidly they can do so only to areas in which they can find the resources they need. Forest dwellers, for example, need trees and could not survive in open grassland. Large grazing animals, on the other hand, control their surroundings by destroying tree seedlings and encouraging the growth of the grasses on which they feed. They might fare better. Overall, however, a rapid warming might well lead to the extinction of many species now living in parks or reserves in the temperate and colder regions unless new reserves could be established and stocked with species transported to them by humans. It would be an enterprise unprecedented both in scale and in kind and while it might be worth the attempt no one knows whether it could succeed, because no one has ever before tried to move entire biological communities in this way.

Crop plants are directly under human control, of course, and can migrate as rapidly as farmers choose. As climatic belts shifted to higher latitudes the regions that are suitable for particular crops would shift with them. In Canada and the Soviet Union the grain-growing belt would extend northward into what is now mainly deciduous forest. Some climate models predict that in lower latitudes the interiors of continents might become drier, increasing the size of the deserts in Central Asia, the central and western United States, southern South America, and southern Africa, but the Sahara might become rather smaller as rainfall increased along its southern boundary.

This prediction could be mistaken, however. Historically, warm periods have been pluvial periods, periods of generally increased rainfall in which all the world's deserts have disappeared. Dry conditions are associated with cold climates and most of all with ice ages. If today the total area of deserts is increasing, other than because of poor land management, it suggests the world climate is becoming cooler rather than warmer, but there is good reason to suppose that in general rainfall is increasing. The one prediction does not preclude the other in the short term, for any climate change is likely to involve fluctuations. A spread of deserts today could prove temporary, however, and give way, perhaps quite quickly, to their reduction in size and eventual disappearance.

In Britain and the rest of north-western Europe the warming may become evident more slowly than in other parts of the world. We are near the northern edge of the zone of temperate climates so we will notice little change as that belt expands northward. At the same time, our climate is influenced by the North Atlantic Drift which will warm only slowly. Eventually, though, we, too, will see changes, provided two assumptions prove correct. We have to assume that the North Atlantic Drift maintains its present course and we have to assume that the northern ice sheets do not melt too rapidly. I have explained the consequences of a change in course of the North Atlantic Drift.

Were the ice sheets to melt too quickly the fresh water released by their melting might float on the surface of the denser salt water, drift far to the south, and so reduce the sea-surface temperature. That would cool the air crossing the Atlantic and either halt the climatic warming or actually reverse it. If the release of fresh water coincided with a cooling caused by a change of course of the North Atlantic Drift, and at an early stage in the warming when temperatures were only slightly higher than they are now, the fresh water might freeze. This would increase the total area of sea ice, and the albedo, and the warming might give way rapidly to a cooling. Under these circumstances our climate would be much colder than it was during the Little Ice Age. We would be unable to grow cereals and our agriculture would be restricted to growing root crops, some hardy vegetables, and the raising of livestock on grass. The farming of red deer might prosper and we might learn from our Scandinavian neighbours and start farming reindeer. It seems an odd kind of prediction to associate with climatic warming, but it is far from impossible and if such cold conditions were to return they could last at least for several centuries.

If our climate becomes warmer rather than colder that will also lead to major agricultural changes. Whether or not we will be able to grow wheat more widely in the north of Scotland will depend on the rainfall – at present it is the high rainfall that prevents wheat from ripening in much of south-western Britain. Maize will ripen further north, perhaps as far north as Scotland, grapes will ripen reliably out of doors at any rate over most of England, and we might be able to grow citrus fruits and olives in more favoured areas.

Rising Sea Levels

This sounds pleasant enough, but there will be another consequence of climatic warming. The sea level will rise. This will increase erosion along low-lying coasts and some areas are likely to be inundated. The map of Britain will change, and so will those of many other countries. This would cause major problems and a glance at an existing atlas will show why. London, Manchester, Liverpool, and Glasgow are among the British cities that might be at risk of flooding – the Thames Barrier is designed to cope with the rise in sea level due to the general isostatic settling of south-eastern England, and not with the additional rise due to climatic warming. Much of the Atlantic seaboard of the United States, from Boston through New York and Philadelphia all the way south to Miami and round the Gulf of Mexico past New Orleans to Houston, lies close to sea level and already there are restrictions on development in many coastal areas that are eroding. An even larger proportion of the total area of The Netherlands is actually below sea level.

Rising sea levels will have another effect, less startling but also serious. Salt water may invade and contaminate the ground water. When it rains, or snow melts, over land some of the water evaporates from the surface and some flows away quite quickly into the nearest river. The remainder soaks down through the soil until it meets a layer of impermeable clay or rock and can sink no further. This is the ground water and it flows, eventually into lakes or rivers, but much more slowly than surface water, so it exists in a region of saturated soil. The upper boundary of the ground water is called the water table and if you dig a hole to below this level the bottom of the hole will fill with water. If the sea level rises, the water table on dry land will rise by the same amount – because water always finds its own level. The amount of ground water could increase, however, only if it were augmented by sea water, and so salt water would contaminate fresh water, making it brackish, and the amount of salt it contained would be likely to increase every time a spell of dry weather reduced the amount of fresh water draining down through the soil.

Growing plants rely on the ground water because when the upper soil is dry, ground water is drawn upward into it by capillary attraction. If the ground water is contaminated by salt water, plants that cannot tolerate salt – and this includes most crop plants – will be poisoned. Water evaporates from the upper layers of the soil, but it evaporates as fresh water that leaves behind as precipitates any salts dissolved in it – remember the experiment with freezing brine. This can cause salination, the accumulation of salt in the surface layers of the soil. When it rains the salts dissolve and some washes away but some drains down to the ground water. Once such a process begins it may be extremely difficult and expensive to reverse it.

Fearing widespread salt-contamination of their farm lands, the Dutch have already taken steps to control the situation. They have deliberately flooded selected areas of land reclaimed from the sea, but flooded them with fresh water. This has created new lakes which have a conservation and amenity value, but it has also raised the water table to above ground level and provided a mass of fresh water that is sufficiently large to exert a countervailing pressure against the encroachment of salt water below the bed. By sacrificing some cultivable land they hope they have protected much more.

In other parts of the world people may be less fortunate. Much of Bangladesh, for example, with a population of more than 100 million, is almost at sea level. Then there are those who live beside river deltas and depend on their waters for irrigation and periodic flooding for the rich silt the floods deposit on their fields. Nearly 48 million people live in Upper and Lower Egypt, close to the banks of the Nile, and millions more along the lower reaches of the Yangtze.

Some land will disappear, and some lands - entire countries - may disappear. There are rather more than 200,000 people living on 197 islands that form part of a group of more than 1,000 islands in the Indian Ocean. The islands are called the Maldives and the local language gave us our word 'atoll' which describes these low-lying coral islands. A small rise in sea level would swamp the Maldives. In Kiribati 68,000 people, nearly 9,000 in Tuvalu, and perhaps 20,000 in the Marshall Islands would suffer a similar fate.

People whose homes and fields disappear will have to find somewhere else to live, but where can they go? In his 1989 Annual Lecture to the Natural Environment Research Council, Sir Crispin Tickell, British ambassador to the United Nations, suggested that if only 1 per cent of the world population – expected to be six billion in the year 2000 – were forced to migrate because of climate change, this would amount to 60 million refugees. The world has never faced a human migration on this scale. In 1988 the UN knew of the existence of some 14 million refugees – with which the nations of the world can hardly be said to be coping adequately – and so we are contemplating a more than four-fold increase.

We should not underestimate the difficulties we would face. It is not simply a matter of educating people out of their racist attitudes so they will accept as neighbours people who look and sound different from themselves – though the chances of doing that are not high. The refugees themselves are likely to represent the poorest and least educated sections of their own societies – almost by definition at once the most vulnerable and the least adaptable. We must hope we can assume sufficient moral stature to address this human crisis and at the same time we must hope that its victims can survive such resettlement as we can arrange into a world that will seem to them more alien than anything you or I can imagine. Even if they can be resettled our experience with refugees suggests that will not be the end of the problems. Their survival will depend on the social services or private charity of the host communities that accept them, perhaps for the remainder of their lives, for refugees are seldom assimilated into the societies in which they are resettled and if they aggregate in large numbers their poverty is likely to affect the surrounding area.

Should the greenhouse warming take place, therefore, its effects will be mixed. World and regional food production is unlikely to suffer because the agriculturally-productive areas will merely shift and may well increase. Living conditions for most people may actually improve, unless the general warming is accompanied by a local cooling of the kind I have suggested might occur in north-western Europe. If not, then a Britain in which temperatures approximate to those of the northern part of the present-day Mediterranean region sounds appealing.

These would be the improvements but there would also be deteriorations whose first and most obvious implication is their expense. The rate of climatic change could present a serious risk of the extinction of many natural but presently protected plant and animal communities and perhaps of species. It might be possible to minimize this loss by establishing new parks, reserves, and sanctuaries, but the cost would be high and it would be only one of the expenses we might feel obliged to meet.

Rising sea levels would threaten coastlines and coastal communities. The advance of the sea might be halted, but only at the cost of constructing defences on an unprecedented scale. We might feel that in some places such costs had to be borne, for fear of losing some of our major cities. Finally the world community might face a situation in which millions of people lost their homes, and some their homelands. What would it cost us to find new homes for them?

Attractive though the idea of growing oranges and olives in your back garden may seem, perhaps you feel that the price is too high. It is too late to prevent such climatic consequences as may attend the quantities of greenhouse gases we have already released, for they are in the air now, but it might be possible to limit future releases and you may feel that we should do whatever we can to this end.

Preventive measures will not be cheap, either, and, as I shall explain in the next chapter, they will demand of us a realism in our view of the world that will shatter certain popular illusions. We face hard choices.

Or do we? Might the greenhouse effect be no more than a mirage, created by computers and existing only in the crude models of the world they produce, the various computers being in general agreement with one another for the very good reason that the scientists operating them feed them the same equations and sets of assumptions? So far, after all, there is no conclusive evidence that a greenhouse warming has begun. That it will begin is calculation, estimate, not indisputable fact based on observation.

We are short of facts but well supplied with apocalyptic prophesies and whatever we do, even if we do nothing, we run a risk. If we take remedial action unnecessarily - in which case will we ever discover it was unnecessary? - we incur heavy costs and great disturbance. If we take no action or inappropriate action, and the greenhouse effect is real, we also incur heavy costs and great disturbance. Our decision - and it is ours - must turn on our evaluation of what an accountant might call the relative costs and benefits. It is time, therefore, to look at the measures that are available to us.

CHAPTER 8

What Can We Do About It?

We belong to a culture that prides itself on its dynamism, on its genius for acting decisively and quickly. Faced with the grave implications of a rapid change in the world's climate it is only natural that we feel impelled to do something about it. We believe we must take action, and do it now.

Assessing the Situation

So what action shall we take? Obviously, this depends on the situation with which we are attempting to deal. What is that situation?

Is it a rapid and substantial climatic warming? The prediction that this is the threat we face is based on computer models that agree with one another but that are very crude. What if green plants, especially those single-celled plants that drift near the surface of the sea, are able to absorb the carbon dioxide we release? What if emissions from these marine plants provide the condensation nuclei that allow cloudiness to increase over the oceans, so shading them and offsetting the warming? What if the combined effects of these natural responses means the climate remains unchanged? If we take action to prevent a warming that does not occur, might we trigger a cooling?

What if the natural responses to an increase in atmospheric greenhouse gases lead to an over-correction, so the warming gives way suddenly and rapidly to a cooling, perhaps even to the full ice age that is now due?

What if the warming happens, but then accelerates because of a positive feedback? This might happen, for example, if the rate of carbon dioxide absorption decreased in the warmer oceans or the melting of the permafrost released a great surge of carbon dioxide and methane. Were this to happen our efforts to limit the change might be ineffectual and it might be more profitable for us to accept the inevitability of change and plan how we would live with it.

Would this necessarily be a bad thing? Might a warmer world not be a pleasanter, more hospitable world in which human life was easier once the initial disruptions were overcome?

With this degree of doubt, should we not hesitate before taking any action at all? After all, when it comes down to it the only thing we really know, and that only in theory, is that the world climate has become unstable. At present there is no conclusive evidence that the temperature is changing as a result of human activities and it may be one or even two decades before we know for certain. It may seem sensible to wait, but if we do is there not a serious risk that by the time we are convinced by the evidence it will be too late to do anything about it because the change we detect is by then well advanced and irreversible?

Clearly, we are not going to find simple, straightforward answers and those who urge us to take remedial action immediately are not being very helpful. The steps they recommend may seem appropriate, but if they are based on one particular scenario - a warming of a certain amount affecting certain regions by a certain date, for example - they might even make matters worse if the change turned out to be quite different. It may seem unduly conservative, even complacent, to warn against rushing to adopt instant solutions, but for the present it is best to be cautious. So what, if anything, shall we do?

More Research; Watching for the Signs

There is one thing we can and must do. We must find out what is happening and what is going to happen. This means ensuring that the relevant scientific research is provided with the resources it needs. In the Britain of 1989 research is not being funded adequately and morale among scientists has never been lower. This is especially true of environmental scientists in academic institutions. I have been told of a research ship, which could have been out at sea measuring changes, that had to remain tied up in port because the laboratory could not afford fuel for it. This is a disgraceful situation that must be remedied.

That said, the research we need will involve workers from many fields and it will not be cheap in the short term – although it may save us from much greater expense later. The computer models need to be improved and refined. To achieve this we will need to provide more supercomputers for the modellers in order to increase the number and variety of GCMs. Models that include slightly different configurations, slightly different assumptions, and that model slightly different types and degrees of change could provide a range of scenarios from which we could build up a more complete picture of our possible futures. At the same time we should foster the development of still faster and more powerful computers that can provide a finer resolution - grids whose sides are kilometres or tens of kilometres long rather than hundreds - and that can deal much more fully with cloud formation and precipitation and with the relationship between the sea-surface temperature and the atmosphere. We will not be able to make fully informed decisions until we know, as reliably as possible, how each region of the world will be affected.

The modellers will need information to feed into their computers and so we should intensify our routine monitoring of what is happening. There are obvious ways to monitor the global environment - by satellite surveillance, for example - and other, more modest, but equally useful ways. Temperatures are measured by weather stations, ships and aircraft, but weather stations are rather few and scattered, especially in the southern hemisphere, and ships and aircraft travel in large numbers but only along certain routes, leaving huge areas in which observations are rarely made. More observations will supply more raw data and this will be of value only if computers and workers to operate them are available to process the data and extract meaning from them.

If you think of the range of temperatures that occur every day and night throughout the world, and the considerable difference there can be between ordinarily warm and ordinarily cool weather, you will appreciate how difficult it is to detect very small changes, perhaps of one degree or even less, against the background noise. On land there are other ways of monitoring climate change, however, and Britain is geographically well placed to make a major contribution to this task. We are very close to the limit of the range of many plants, especially trees, and quite small changes in temperature will affect their behaviour.

Lime or linden (*Tilia*) rarely sets viable seed in Britain because conditions are too cold for it. Should it begin to do so and should the range within which it sets seed extend northward, this would indicate a general warming.

Hazel (*Corylus avellana*) also sets seed - its nuts - unreliably and there were fears that both it and oak (*Quercus*) were declining in Britain and might become extinct, presumably because of fears that the cooling which occurred some years ago might continue. More hazel nuts and acorns, and more good years for them, might also indicate a warming.

The spread of holly (*Ilex aquifolium*), which cannot tolerate too many really hard frosts, would be an even better indicator. It is a broadleaved evergreen, a type of tree that is more typical of southern climates than of Britain, and it has always been of particular interest to students of past climates.

I see no reason why amateur botanists should not play a very useful part in this job of tree-spotting. The species concerned are few in number and easy to identify and if every enthusiast were to make a point of watching out for the appearance of new seedlings close to home the entire country could be monitored continually, thoroughly and cheaply - provided, that is, there were receiving centres, staffed by professionals, to accept the reports and process them to extract information from them.

Having found employment for amateur botanists it would be almost churlish not to do as much for amateur entomologists, for the insects, too, will respond to climate change. Britain is close to the climatic limit for many insects, including butterflies and moths, and the disappearance of butterflies from time to time has much more to do with cold weather than with anything farmers do by way of pesticide spraying.

While the lepidopterists watch the butterflies and moths, other entomologists can usefully keep an eye on what goes on in their own gardens. I mentioned the aphids earlier. Serious aphid infestations very early in the season suggest that adults have overwintered, and the discovery of aphids during the winter would confirm it. Such evidence could be sought from the Midlands northwards.

My friend Lawrence D. Hills, the horticulturist and gardening writer who founded the Henry Doubleday Research Association, which encourages amateur gardeners to undertake this kind of local, low-key, but widespread investigation, has suggested two other pests that might bear watching. The cabbage whitefly (*Aleyrodes proletella*) is really a tropical species, as are all whiteflies, and at the very limit of its range in southern England. It is rare north of Nottingham but if it begins to appear much further north, or becomes more widespread, and if it starts to produce a larger number of generations in a year, it would mean the climate has grown warmer.

The other pest is the red spider mite (*Tetranchus urticae*), curiously named because it is more a brownish orange than red and it is a mite but not a spider - although it looks a little like a spider and it spins a web of sorts. It is a pest almost exclusively of glasshouses because its development is very slow at the temperatures it encounters outdoors in Britain. At 10°C (50°F) it takes two months for it to develop from egg to adult but as temperatures rise its growth accelerates. It occurs outdoors in sheltered parts of Cornwall, where it is a pest of such crops as strawberries and courgettes. Should infestations start to occur on outdoor crops in other parts of the country that, too, would be an indication of warmer conditions.

Immediate Action

It is all very well and necessary to watch for early signs of change and to reinforce our efforts to understand what may be happening, but what shall we do if we discover the predictions are correct and the climate is changing? Then we might feel compelled to act and so it would make sense to start now in planning a range of responses. It will take time. Anything that involves altering the way we live or the way our industries operate cannot be done quickly. It must be planned and then implemented in a controlled way and no matter how urgent the situation we diagnose, undue haste will create far more problems than it solves.

Whether we face warmer conditions or cooler ones there is no doubt that our release of greenhouse gases threatens to destabilize the world climate. Nor can we doubt that destabilization is dangerous. The system with which we are interfering is very large, driven by forces far beyond our capacity to match, and our present understanding of the way it works is rudimentary. The position was summed up in the final statement released in Toronto in June, 1988, at the end of the International Conference on the Changing Atmosphere: Implications for Global Security. Part of the statement said: 'Humanity is conducting an unintended, uncontrolled, globally pervasive experiment whose ultimate consequences could be second only to a global nuclear war'. Our plans, therefore, must be aimed primarily at reducing our level of interference and this means cutting down our emissions of greenhouse gases.

We who live in the older industrialized countries have a special responsibility. At present, of course, we emit far more greenhouse gases than the less industrialized countries. That alone obliges us to mend our ways if we can, but it is not the only reason. In years to come other countries will increase their industrial output. Not only is this inevitable, it is desirable. The dream some environmentalists have of bringing economic growth to an end is quite unattainable. This is fortunate, for its realization would be environmentally catastrophic. Throughout most of the world it is only by some form of industrialization that standards of living can be improved, and standards of living must improve if birth rates are to be brought into line with death rates, eventually to stabilize population sizes.

The way to control over-rapid population growth has been well known for many years. It requires mainly that parents be given access to education and then to remunerative and satisfying employment, and that old people be provided for adequately at public expense. This removes the need for children to provide cheap labour and to be guardians and providers to their ageing parents, while at the same time giving working parents a financial incentive to postpone having children. Over a few generations this stabilizes the size of populations and it also reduces poverty. Reducing poverty reduces the pressure on such environmentally sensitive resources as marginal land and forests.

Development in the less industrialized countries is necessary but this will mean an inevitable increase in their consumption of fuels. We must allow them this increase and we must accept that because for some years they will remain short of capital and technical expertise they will have to utilize the simpler, more easily accessible technologies. They will have to burn carbon-based fuels. If they are to do so, and the global output of carbon dioxide is still to be restricted, the rich countries will have to make disproportionately large reductions in their own consumption of such fuels.

There are various things we might consider doing in pursuit of this goal and although none is easy some of them would bring incidental benefits, almost as side effects. You might say that our present experiment with the atmosphere can be traced back to our failure to take proper account of the side effects of our activities, so it would be foolish to ignore the side effects of our remedies. Where those side effects are beneficial in their own right, however, we may find ourselves in the curious position of favouring them for that reason alone – they are sensible whether or not the climate changes.

Energy Conservation

Obviously, we can reduce the amount of carbon dioxide we release if we burn less carbon-based fuel. We can do that either by reducing our consumption of fuel or by moving to sources of energy that are not derived from carbon, or by some combination of both. It is the reduction in fuel consumption, called energy conservation, that attracts most widespread support from environmentalists.

When I was a boy you could travel from just about anywhere to anywhere else in Britain by a combination of train and bus. Over the years since then the political dogmatism that favours the private sector of the economy at the expense of the public sector has allowed the emphasis in transport to shift profoundly. Today, rail services except those between principal cities and some commuter lines are sparse, mainline services are expensive, and the trains themselves are often overcrowded and uncomfortable. The deregulation of bus services has come close to abolishing time tables because only one week's notice is required of a change of service. The predictable result is that most of us who own cars use them in preference to public transport. Road vehicles are a major source of carbon dioxide.

It is not true that public transport is necessarily more efficient than private transport in energy terms, but it could be made so. A car carrying four people uses less fuel per passenger-mile than a half-filled train or bus, but a more comprehensive public transport network and lower fares might attract passengers who would otherwise travel by road. This could lead to an energy saving and in terms of fuel use there is another benefit, at least in the case of trains. With the electrification of rail services, locomotives are not committed to the burning of carbon-based fuel. They use electricity, and this can be generated without releasing any greenhouse gases at all if only we can bring ourselves to accept that nuclear power is environmentally preferable to the burning of coal, oil, or gas.

A shift in emphasis from private to public transport would be worthwhile regardless of climate change. It would ease the congestion on roads and increase the speed of travel in most of our cities, while reducing local air pollution in city centres. These are environmental benefits and there would also be social benefits. Those of us who own cars forget too easily that even today there are many people who do not. Young people for example, are either too young to drive or too poor to own cars; but housewives, staying at home while their husbands use the family car to travel to work, comprise the largest single group of victims of our transport policy.

Such a shift would not forbid the use of private cars, far less abolish them. So can cars be made more efficient? Perhaps they can. Great advances have been made in recent years and more may be possible, although we cannot be certain of this. Eventually, though, it may be possible to substitute alternatives for the present generation of petrol and diesel engines – engines that do not depend on carbon-based fuels. The most likely candidates are electric cars, powered by batteries, cars that use hydrogen as a fuel, and cars that use methanol.

It is true that methanol (or, less satisfactorily, ethanol) is a carbon-based fuel, whose combustion releases carbon dioxide; but it is obtained from agricultural crops grown specifically for the purpose or from farm wastes. The growing of the plants absorbs precisely the same amount of carbon dioxide that the burning of the fuel emits, so methanol-fuelled engines add no carbon dioxide to the atmosphere overall. Such engines are not without their problems but it is possible that these might be overcome. There is no space here to describe in detail how they might work, or how well the vehicles using them might perform, but if you wish to learn more they are described fully in an article, 'The Case for Methanol', by Charles L. Gray, Jr. and Jeffrey A. Alson, in the November 1989, edition of *Scientific American*.

Of course, crops grown for fuel must use land that might otherwise be growing food, and so the criticisms of cash-cropping in Third World countries must apply equally to fuel crops. On the other hand, were we to move substantially from petroleum to methanol, making methanol a valuable commodity, the economic benefits of our appetite for fuel might be spread more widely among the less-developed countries than they are today. The case can be argued both ways.

Electric cars, like electric trains, emit no pollutants of any kind and the electricity to charge their batteries can be generated cleanly. The difficulty arises from the weight of the batteries. These are so heavy that much of the work the engine does is devoted to carrying them and so far the performance of electric cars does not make them attractive – their top speed is low and their range short. Much research effort is being directed toward the development of lighter and more powerful batteries. When and if it succeeds, electric cars may become a realistic proposition.

Some years ago there was talk of the hybrid car that would be a compromise between an ordinary, petrol-burning car and an electric car. The hybrid car had a small conventional engine that was used only to sustain the charge on the batteries which supplied the motive power for the vehicle. This increased the range of the car, though not its speed, while using little fuel because the petrol engine could run constantly at its most economic speed.

Hydrogen presents fewer problems, provided it can be obtained with less expenditure of energy than is produced by burning it. The most promising way to obtain it is probably by the electrolysis of water, a process that breaks one of the bonds holding the water molecule together, so that H—O—H (another way of writing H_2O) becomes H + O—H. The development of photovoltaic cells, that convert sunlight into electrical energy, has reached a stage where it may soon be feasible to use them to produce hydrogen. The hydrogen will not be free because although the sunlight is free the cells are not, but the process involves no extra expenditure of energy and there are no by-products of any kind.

The disadvantages of hydrogen arise from the volume of it that must be carried and the fact that at ordinary temperatures it is a gas, not a liquid, but both have been largely overcome. An experimental Mercedes car has been built that looks much like any other Mercedes but that runs on hydrogen. The gas forms a loose compound with certain metals, so the fuel tank is filled with metal, in powdered form to increase its total surface area, and charged with hydrogen gas which it then holds. The gas is released when the mixture is warmed a little. Hydrogen is not quite problem-free, even so. When it burns the only product is water vapour - the reaction oxidizes the hydrogen as with any true burning - and water vapour is a greenhouse gas.

People still have a fear of hydrogen based on the serious fires that destroyed several airships in the 1930s, but hydrogen is no more dangerous than most other fuels and probably less so. In an accidental fire, hydrogen, unlike all petroleum fuels, does not scatter blazing drops because it is a gas, so the fire spreads less readily than other fuel fires and causes fewer injuries. Being light, a hydrogen fire burns upwards, the burning gas rising rapidly, away from people and structures on the ground.

Improving the efficiency of cars is of obvious benefit to motorists because it reduces their running costs. It is too early to say how soon, or even whether, alternatives to existing engines will offer a similar advantage.

Saving energy in the home will also reduce fuel bills. If such domestic machines as cookers, washing machines, refrigerators and freezers can be made to work satisfactorily while using less electricity or gas, clearly it would be a good idea to introduce them.

The insulation of buildings might also be improved, although there are probably few gains to be made here.

Good insulation is now routinely incorporated into new buildings, and many old buildings which could be were insulated during the energy efficiency drives of the 1970s. This also applies to industrial and commercial buildings. The efficiency of some industrial machines and processes might be improved but the decline in British manufacturing has eliminated or greatly reduced many of the more energy-intensive heavy industries.

While energy conservation makes obvious sense the measures it involves cannot be introduced quickly without incurring high energy costs of their own. A change in emphasis from private to public transport may mean Britain needs to plan rather fewer roads for the future - although cars that already exist must be accommodated - but it will mean building more buses and railway rolling stock, and laying more railway lines, all of them activities that consume fuel. It will take quite a number of years for new types of cars, marketed as they are developed, to replace all the existing cars. New buildings are more energy-efficient than old ones, but we can hardly attack every building more than about 20 years old in a manic programme of demolition and replacement.

Coal-Burning Power Plants

There is talk of encouraging conservation by imposing a 'polluting tax' on products, such as carbon-based fuels, that are believed to be environmentally harmful. I doubt whether this will do much more than raise revenue unless it is accompanied by substantial public investment in other reforms. Until public transport services are improved people will continue to use their cars because they have no alternative, and taxation cannot encourage the use of entirely new types of car until such cars become available. If the tax is levied honestly, however, it could in time alter the mix of methods for electricity generation by imposing on coal-burning plants emission limits equivalent to those applied to nuclear plants and so making coal-burning much the more expensive of the two.

Large installations, with well-designed and well-maintained equipment and subject to controls that can be enforced, are able to burn fuel more efficiently - which means more cleanly - than most small installations. Until the Smoke Abatement Act restricted them, domestic fires were one of Britain's most serious sources of urban air pollution and although a change to smokeless solid fuel eliminates the emission of black smoke, there is no reduction in the emission of other pollutants. Devices that burn fuel slowly at relatively low temperatures use their fuel economically but are no less polluting, and wood-burning stoves are perhaps the most polluting devices ever invented. If we wish to control air pollution generally we must come to rely more heavily on electricity, whose use causes no pollution at all, generated as efficiently as possible in well-designed and well-managed power stations.

Most conventional, fuel-burning or nuclear power stations actually generate electricity by heating water to produce steam and using the steam to spin turbines. This is effective, but it could be made more efficient. When it leaves the turbines the steam is recycled to avoid having to waste fuel heating cold water, but in some power stations surplus heat is piped as hot water to nearby buildings. Depending on the distance between the power station and its neighbours, this water can be used to heat glasshouses or to provide space heating for factories or even homes. The installation is expensive, not every power station is suitably sited, and the overall energy saving may be smaller than is sometimes supposed, but the technique can provide useful savings in some places.

Power generation itself, an even greater source of greenhouse gas emissions than cars, provides more opportunities for reform without requiring substantial changes in the way people live. However attractive the more extreme environmentalists find it, large-scale social engineering - which is what most green policies imply - is difficult and its outcome very uncertain. The gradual replacement of one type of power station by another is a great deal simpler.

It has been suggested that existing oil- and coal-fired stations be converted to burn natural gas (methane). This probably would not be difficult and it would reduce carbon dioxide emissions because methane can be oxidized to release heat much more efficiently than either of the other fuels, so less of it is required to produce each unit of heat. Such a solution would be only temporary, of course, because reserves of natural gas are not limitless and a major increase in our use of them will deplete them more quickly. Nor is it a complete solution, because efficient though methane may be, it is still a carbon-based fuel and the by-products of its combustion are the greenhouse gases carbon dioxide and water vapour. Nevertheless, for what remains of their working lives the conversion of existing power stations would certainly help. After that they will have to be replaced by plants using quite different fuels or converting energy in different ways.

Renewables

It is possible for buildings to supply some of their own energy. Solar power may contribute a little, but only a little. Passive solar collectors are simple and based on familiar technology. Fitted to the roof or exterior walls of a building they consist of shallow, tray-like boxes with an exposed surface of a heat-absorbing material below which a closed system of piping carries water. The water is warmed by the absorbed heat and carried indoors where it warms the water in the hot water tank. The device works well in summer but less well in winter. Solar cells, the photovoltaic devices I mentioned earlier, use the light rather than the heat of the Sun and so work as well in winter as in summer – apart from the fact that in high latitudes hours of daylight are reduced in winter. A large area must be covered with them if useful amounts of power are to be produced, and at present the power is expensive, but prices are falling and the day may come when large buildings – though probably not private houses – are able to install them profitably.

Heat pumps and fuel cells may also make useful contributions. A heat pump circulates a fluid between the inside and outside of a building. When it is outside the fluid vaporizes, absorbing its latent heat of vaporization. Inside the building a compressor causes it to condense, releasing the same amount of latent heat. In effect the heat pump takes energy from a cold area and releases it in a warm area – rather like a refrigerator running backwards.

A fuel cell is essentially a box with an electrode at each end. Oxygen is passed across one electrode, hydrogen or some other fuel across the other, but the oxygen and the fuel do not come into contact with one another. Electrons are given up by the fuel and accepted by the oxygen and a flow of electrons - an electric current - is established between the electrodes. Fuel cells are silent, clean, and very efficient, but may not be commercially available for some years.

Wind power can provide useful energy in small amounts to farms and remote communities that are beyond the reach of more conventional supplies. In some places it already fills this role. Any idea that wind generators could replace any - far less all - British power stations, however, is pure fantasy. 'Wind farms' have been built in some countries, such as the United States, where desert sites are available, but it is doubtful whether suitable sites could be found anywhere in Europe. The environmental objections would be formidable.

A modern wind generator is more than 40 metres (130 feet) tall, has two or three blades more than 18 metres (60 feet) long, and generates between 1 and 1.5 megawatts of power. Between 650 and 1,000 such devices are needed to produce as much power as one modern power station of modest capacity and more if they are to equal the output of a large power station. At the Altamont Pass wind farm in California, for example, 300 wind generators produce one-quarter the power of a single power station. The generators cannot stand too close together because each would interfere with the wind flow around its neighbours and in practice a wind-powered alternative to an ordinary power station would occupy not less than, and probably much more than, about 1,200 hectares (3,000 acres) for the generators alone. In addition there would be buildings housing transformers and grid pylons to carry the electricity into the public supply. The total installation would be extremely large and to ensure the reliability of the wind it would have to be sited on coastal cliffs or an upland moor, and well away from any town or village because the generators are very noisy. It is possible that the land around the generators might be grassed and used to graze livestock, but apart from farmers and their employees the public would have to be excluded on safety grounds. Should a turbine fail it could cause serious injuries.

The vertical motion of sea waves has also been studied as a source of generating power, based either on devices that float at the surface and rock back and forth, or on cylinders fixed to the sea bed in which a column of water oscillates as the waves rise and fall. Both types of device require very large installations if they are to produce useful amounts of electricity and they have to be capable of withstanding severe winter storms. An experimental Norwegian oscillating-column device worked well and aroused much interest a year or two ago, until a storm demolished it one dark winter night.

Tidal barrages have also been considered but they are very expensive and there is a shortage of sites in which the range of tidal movement is large enough to make a barrage worthwhile. The most suitable sites are usually estuaries. The barrage is actually a dam. The ebb, and in some designs also the flow, of the tide turns turbines in the dam and the turbines generate electricity. Again, there are grave environmental objections because any large-scale interference with the flow of water in an estuary is likely to alter patterns of sedimentation. This could profoundly affect the wildlife of the area and possibly commercial fisheries, for many fish breed or spend part of their lives in estuarine waters.

The movement of water can generate power, of course, and does so in hydroelectric schemes. These have been in operation for many years and in Britain all the suitable sites that can be exploited have been.

Geothermal energy also has little to offer. It exploits local reservoirs of water, or in Britain dry rock, deep below ground that are hotter than the rock surrounding them. The temperature rises with increasing depth in all rock, because of the decay of radioactive elements in the rocks, and the anomalously hot rocks therefore offer a version of nuclear power.

In principle, two holes are drilled into the reservoir and in the case of hot dry rock the rock between the holes is shattered to make it more permeable. Cold water is pumped under pressure down one hole and returns, heated, up the other. The water can then be used directly, if it is hot enough, or can be heated further but with less fuel than would otherwise be necessary. The water that has passed through the rock must be kept isolated from the environment and must not be allowed to contaminate water intended for domestic or industrial use because it is likely to contain substances dissolved from the rocks. It may be very corrosive, for example.

The energy must be used close to the geothermal site, because hot water cools if it is transported far, and such sites are not necessarily close to urban areas where energy is needed. Nor is it inexhaustible. Passing cold water through hot rock cools the rock and eventually - probably in about 30 years - the once hot rock has been cooled to the same temperature as its surroundings and no more energy can be extracted from it.

The so-called renewable sources of energy sound attractive but there are serious environmental objections to all of them, not to mention their rather high cost. There are also grave doubts about their reliability. There really is no convenient substitute, all ready and waiting to be installed, for conventional power stations.

Those environmentalists who suggest there has been some kind of conspiracy to keep existing types of power station in operation when ample supplies of clean, 'natural' energy were to be had for the asking are quite mistaken. Nor is it true that research into renewables has been seriously starved of funds. They have been studied very thoroughly indeed over many years - research into wind power, for example, dates from the 1940s. Geothermal energy in the form of hot springs has been used from ancient times, but hot springs occur in only a few places. The extraction of energy from hot dry rock, as in Cornwall, is expensive, mainly because it requires the drilling of boreholes deep into hard rock, but it is being funded adequately. The other renewables involve devices that individually are not very costly and so the funds needed to build and test them are relatively small, and many such devices have been built over the years. The cost rises when full-scale demonstration versions have to be installed, but in the case of wind power such demonstration installations exist in several countries, and are being planned for Britain; and wave-power machines have been operated, though not on a full scale, in Britain. Information on tidal-barrage devices is derived partly from the installation in the Rance estuary, in Britanny, which has been in operation for many years, and partly from theoretical studies whose main purpose is to locate the best sites and designs and, to a lesser extent, to consider the environmental implications. You cannot build a small experimental or demonstration model of a tidal barrage; it has to be built to full size from the start. The fact is that enough is known about renewables for informed judgements to be made about them.

This is not to say the renewables have no contribution to make, only that the contribution is small, amounting in most estimates to no more than a few per cent of our total energy needs. The most optimistic estimates for Britain, assuming local and environmental objections can be overcome, is that eventually they might contribute up to 10 per cent of the country's total electrical power.

Policy Options

It is pointless to pretend we do not need substantial amounts of energy. Conservation may reduce our demand but it will not end it, for it is our use of energy that underpins our entire way of life. Extreme environmentalists may regard that way of life as undesirable and may view its ending with equanimity, or even pleasurable anticipation, but they offer no realistic alternative and perhaps have taken too little account of the environmental impact of large numbers of impoverished people burning wood and other plant material for want of electricity.

Since we must have power stations the choice is between types. We will have to accept an increase in the proportion of the energy we use that reaches us in the form of electricity. This may not imply more power stations if energy conservation can make a significant contribution, but if that energy is to be generated with the least pollution and with no emission of carbon dioxide we will have to accept a very substantial increase in the proportion of our electricity that is generated in nuclear power stations. Environmentalists have made this a contentious issue and have aggravated a genuine public anxiety about nuclear power until it has become a wholly irrational fear. It is a matter I shall discuss in the final chapter.

It is much easier to imagine ways in which we might reduce our mainly industrial emissions of carbon dioxide than it is to devise an effective means for halting or even regulating the clearance of forests in the humid tropics. Exhortations from European and American scientists, environmentalists, and even politicians have little effect and may even be counter-productive. People do not like being told by foreigners how they should manage what they regard as their own affairs. In their position we would take a similar view. There is force, too, in the argument that we cleared most of our own forests centuries ago and have no right to prevent others from pursuing a similar developmental path.

The truth is that the governments of the countries concerned do not permit the exploitation of their forests from mere whim. They feel constrained to earn foreign exchange in any way they can, not least in order to pay some of the interest on their loans from Western banks, and they believe they can alleviate the problems of their landless poor by encouraging them to settle the once forested land. This policy is probably mistaken but it is what the governments believe and it explains why environmentalist schemes to impose international bans on trading in tropical forest products and to halt development projects that involve forest clearance have not the slightest hope of success. They are rather like advising a starving man that he could be wealthier if he ate less. The difficulties of trying to impose a moratorium on commercial whaling and to regulate the international trade in endangered species should warn us against embarking on the probably impossible task of banning trade in forest products.

Within the last year or so a small number of suggestions have been made for ways of exchanging debts for protection of the forests and the funding of some development projects has been tied to conservation programmes, with even the World Bank becoming involved. Some of the new ideas are imaginative, perhaps the most imaginative being that of Sir James Goldsmith. He suggested that a company be formed to buy national debts at their market value - which is less than their face value - on condition that the sums paid be regarded as rent for areas of forest that would then be protected against inappropriate development. This seems to overlook the need for internal political and social reform in some tropical countries and partly for this reason such ideas may turn out to be impracticable, but they have the virtue of being positive and we can be encouraged that at last the need to protect tropical forests is being taken seriously by those who might be able to achieve it.

If we economize in our use of carbon-based fuels we will also reduce our emissions of other greenhouse gases. Nitrous oxide and ozone, between them calculated to contribute 18 per cent of the predicted greenhouse warming, are linked to combustion. Gaseous nitrogen is chemically unreactive but at high temperatures and pressures, such as those generated in a modern car engine, it will oxidize. Once in the air, together with unburned hydrocarbons – compounds consisting mainly of hydrogen and carbon – it can engage in further reactions, powered by sunlight, that produce a range of pollutants one of which is ozone.

Action has been taken already to curb the use of CFCs, said to be responsible for 14 per cent of the predicted warming. The industrialized countries have pledged to phase them out completely by the end of the century. Two problems remain. The first is that although the industrialized countries plan to end their use of CFCs, as standards of living improve in the less-developed nations CFC use will increase again, mainly because of increasing demand for refrigerators, freezers and air conditioners. This difficulty can be overcome if full information about the substitutes being developed in the industrialized countries, and the technology needed to manufacture them, can be transferred to developing countries.

The other problem arises from the real possibility that while the substitutes may decompose rapidly before reaching the stratosphere where they might affect the ozone layer during their residence time in the troposphere they may be greenhouse gases no less effective than the CFCs they replace.

The second most important greenhouse gas is methane, large amounts of which are released as a by-product of the digestion of celluloses by bacteria in the guts of ruminant animals and in the muds of rice paddies. We can do nothing that would reduce the output of paddy rice. It is a staple food and essential.

Can we reduce the number of ruminant animals, especially of cattle since it is their numbers that have increased? In the rich, industrialized countries where most meat is eaten their products are not nutritionally essential. Indeed, some are believed to be harmful in the quantities we consume. We would suffer no more than mild inconvenience were there to be fewer cattle in the world.

Evolutionarily, cattle are browsers and grazers. In modern times they have most usually been raised in fields and fed mainly with grass, the diet of dairy cattle being augmented with protein supplements to increase milk yield. In Britain beef is still almost wholly grass-fed, but Britain is fortunate in having ample pasturage. Our climate and soils are good for growing grass. In most other countries this is not the case and cattle are raised not on grass but on cereal grains and protein supplements. This allows the food to be taken to the animals, rather than leaving the animals to find their own, and it makes possible the crowding of large numbers of livestock into an area far smaller than would be required if the animals were grazing. It is the basis of the feedlot method of cattle husbandry and provided food is obtainable and can be transported and distributed efficiently there seems to be no limit to the size a feedlot unit can attain.

This is a main part of the problem. The link between the livestock and its traditional food supply has been broken. If steps could be taken to re-establish that link, so the number of cattle in a given area – the stocking density – was related directly to the food available for them in that area as it would be if they were grazing, then the number of cattle would fall dramatically.

Farmers would suffer, of course, but perhaps only temporarily for doubtless they would find other foods they could produce. Consumers need not suffer seriously since the cereal foods eaten by the cattle could be eaten by humans – and more economically because it takes about ten pounds of cereals to produce one pound of beef. A diet so dominated by cereals may sound unappetizing, but it is not difficult to extract the protein and process it into meat substitutes.

As I have said, it would be premature to take action specifically to deal with the greenhouse effect until we know much more about it than we do now. I see no reason, though, why we should not plan measures to reduce our emissions of greenhouse gases, and even make a start at implementing them, if they bring recognizable benefits unrelated to the climate.

If we save fuel we will husband resources, reduce various forms of environmental pollution and live more cheaply. Saving fuel makes obvious sense. If someone can market a car I can afford that will cover twice as many miles on a gallon of petrol as my present car I would be foolish not to buy it, as I would be foolish not to use public transport if that were convenient and cheaper than driving.

Such measures as these promise direct, personal benefits. The protection of tropical forests offers benefits that are more subtle. Of course I will pay lip service to the conservation of tropical species, but I have never visited the tropics and it seems unlikely that I ever will. I have no personal link with them. However, I must be touched by measures that promise to improve the lot of the poorest people in the world, as I am outraged by those who advance the false Malthusian argument that assisting the poor merely increases poverty. I have never seen the tropics but I have seen poverty. Its alleviation is a benefit I recognize.

It may seem more difficult to see what incidental benefit might accrue from curbing the more extreme forms of livestock farming, such as feedlot husbandry. Yet that, too, would lead to improvements in the local environment. Feedlots produce very large amounts of manure that can cause serious pollution and are unsightly.

This leaves us with the choice I said we must make among the alternative methods for generating electrical power in conventional power stations. I hold to my argument that unless some incidental benefit results from making such a choice we should postpone it until we know more about the future of our climate. I believe such a benefit would result if we were to make greater use of nuclear power and the time has come for me to explain and justify that belief.

CHAPTER 9

Going Nuclear?

Most countries are increasing their emissions of carbon dioxide, but as I mentioned in chapter 5 there are two notable exceptions. Between 1986 and 1987, carbon dioxide emissions fell by three per cent in France and by two per cent in West Germany. Both countries sought to use energy economically and this accounts for much of the reduction but the remainder is due to the fact that both countries, France especially, rely heavily on nuclear power. Despite the anti-nuclear influence of the West German Greens, a large, new, turbine generator came into service in 1988 at the Brokdorf nuclear power station in Schleswig Holstein. It has a capacity of 1,330 megawatts.

At the beginning of 1988, there were 417 civil nuclear reactors in the world as a whole, in 26 countries, delivering a total of 300 gigawatts (a gigawatt is one thousand million watts) and, despite all the opposition, 23 new reactors were commissioned during 1987. These reactors emit no carbon dioxide, indeed, they emit virtually nothing at all.

Cleaning Up Carbon-Based Fuel

When you burn any fuel containing carbon, carbon dioxide is released. This is unavoidable because it is the oxidation of carbon ($C + O_2 \rightarrow CO_2$) that supplies the energy. The use of natural gas (methane) leads to smaller carbon dioxide emissions but only because the burning of gas yields more energy per tonne than does the burning of any other carbon-based fuel.

Efforts have been made to make the burning of fuels cleaner and most of the products of combustion can be removed from the gases before they are released – at a price. Smoke abatement legislation, for example, forbids the release of black smoke and soot.

Sulphur dioxide is the best known example of a pollutant that can be removed. Fears that this gas may be a principal cause of acid rain, which damages vegetation and causes the acidification of lakes, has led to strong political pressure to install equipment to prevent it from entering the air. The usual technique – called flue gas desulphurization – is to pass the exhaust gases through water containing lime (calcium hydroxide). The sulphur dioxide reacts with the lime to produce calcium sulphate, an insoluble, chemically inert substance whose common name is gypsum.

There is no doubt the technique works, but there are environmental costs and the benefit is doubtful. The calcium hydroxide begins as limestone rock (calcium carbonate), which must be obtained by quarrying. In Britain there are fears that eventually this may lead to a large increase in the amount of quarrying in one of the best sources of limestone – the Peak District National Park. That is the first cost.

The limestone is then kilned – heated strongly to drive off carbon dioxide and so convert it from the carbonate to the oxide ($CaCO_3$ + heat → CaO + CO_2) and the oxide (quicklime) is slaked with water to produce the hydroxide. So the process requires carbon dioxide to be released and that is the second environmental cost.

At the end of the process the gypsum remains for disposal. There are uses for gypsum. It can be processed further to make plaster of Paris or builders' plaster, for example, but in Britain the predicted output from flue gas desulphurization is equal to the present entire national production of gypsum. Either the new supply will find no market, or it will undercut existing prices and so destroy the established industry. Most probably the gypsum will have to be dumped. That is the third cost.

The fourth concerns the removal of the sulphur dioxide itself. In the air it provides nuclei for the condensation of water vapour and cloud formation and some scientists believe the rather slow rate at which the northern hemisphere has been warming may be due to an increase in cloudiness over the years, due in turn to the amount of sulphur dioxide our industries have released. If the climate is growing warmer, ceasing to release sulphur dioxide may accelerate the warming. It would be paradoxical, perhaps even dangerous, to allow one pollutant to offset the effect of another, but there are times when we are compelled to weigh one evil against another and choose the lesser. It is never easy, never comfortable, and I do not recommend this as a practicable course of action; but it is worth pausing a moment to consider just how damaging sulphur dioxide really is.

Sulphur dioxide is implicated in acid rain damage, but it is not the only cause of it and the burning of fuel is not the major source of atmospheric sulphur dioxide in the areas worst affected. Plants are also damaged by nitrogen oxides and especially by ozone, mainly from car exhausts, and some of the forest damage, especially in central Europe, has been caused, singly or in combination, by drought, fungal disease, insect pests, and in some places by bad management. Many lichens are extremely intolerant of airborne sulphur and their presence or absence provides a widely used indicator of sulphur pollution. Years ago, scientists investigating the reported damage in West German forests found sulphur-sensitive lichens growing abundantly, suggesting that sulphur dioxide levels were very low – a finding confirmed later by direct measurements. Where sulphur dioxide is an important pollutant, the source is mainly the emission of dimethyl sulphide by marine algae.

There is a further reason for suspecting that industrial emissions of sulphur dioxide are less serious, at least in Britain, than many people believe. Sulphur dioxide attacks limestone and so it corrodes limestone buildings – of which we have many, located in city centres where pollution levels are likely to be highest. Damage to buildings has not been increasing in recent years in Britain, although there are said to be serious acid pollution problems in Poland, Czechoslovakia and probably in other Eastern European countries.

Sulphur dioxide can be removed but removing carbon dioxide presents difficulties of a different order, although the chemistry is simple enough. As with sulphur dioxide, the process uses water containing lime (calcium hydroxide). The carbon dioxide reacts with the calcium hydroxide to produce calcium carbonate, which is insoluble and can be removed. Disposing of calcium carbonate presents no serious problems since it is a substance that occurs widely in the natural environment as chalk and limestone.

The technology of kilning calcium carbonate to obtain calcium oxide is one of the most ancient we know, and perfectly safe provided care is taken when dealing with the oxide (quicklime). The kilning, however, releases precisely the same amount of carbon dioxide as will be captured ($[CaCO_3 + \text{heat} \rightarrow CaO + CO_2] = [CO_2 + CaO \rightarrow CaCO_3]$), plus any carbon dioxide that may be released by burning fuel to provide the necessary heat. At best the process achieves nothing, at worst it releases more carbon dioxide than it captures.

It might be possible to dispose of carbon dioxide by dissolving it in the sea. This would mean building a pipeline from each power station, or other major carbon-fuel-burning installation, out into the deep ocean. The gas would have to be released in deep water, below the thermocline. If it were pumped into surface waters it would saturate them locally, thus reducing their capacity for absorbing atmospheric carbon dioxide, and eventually the gas might escape from the water into the atmosphere, so defeating the purpose of the operation. It means the pipelines might need to follow the sea bed, extending for more than about two hundred miles beyond the shore. Clearly, this would be very costly and no one knows for certain whether it would work or whether it might cause disturbances at least as serious as those it was meant to remedy.

The Nuclear Alternative

There really is no way around the carbon dioxide and, as I explained in the last chapter, nuclear power is the only feasible alternative to carbon-based fuel if we allow that in years to come, no matter how much energy we conserve and no matter what may happen to our economy, we continue to need electricity. Nuclear power exists now, has existed for the last thirty years, and it will not vanish.

It is true that particular countries, including Britain, could phase out all nuclear power stations if they chose to do so. We could do so even without increasing our reliance on coal, oil, and gas, because there is a two gigawatt cable beneath the Channel linking the British and French electricity grids. We import French power already, and in this situation we would simply need to import more of it. Then, rather than generate our own nuclear power we could use French nuclear power. It hardly seems very sensible.

When the anti-nuclear movement began it was not concerned with reactor safety or with pollution, but with the political implications of the centralization of generating capacity. This, it was argued, placed too much economic power in the hands of those who ran the industry. It is a reasonable objection, but one that applies to large-scale electricity generation of any kind and one that can be answered. At least in theory there is no reason why power stations, including nuclear power stations, cannot be small and owned by the communities in which they are located. Large scale and nuclear generation are not indissolubly linked.

There were, and still are, doubts about the relationship between the civil and military nuclear industries. It is true that the civil industry began as an offshoot of the military need for supplies of materials to make weapons, although the two industries are now separate. The military need continues and will be satisfied regardless of what happens to the civil industry. Closing nuclear power stations would make no difference whatever to the weapons programme because designated military reactors would continue to supply weapons-grade uranium and plutonium.

The more widespread and more recent fears centre on what are perceived as the dangers arising from the use of radioactive materials. They arise partly from the difficulty of understanding this highly complicated technology and partly from the extraordinary precautions that are taken to limit human exposure to radiation. These fears have combined and evolved into an entire mythology surrounding words like 'nuclear' and 'radiation' and many environmental groups describe every incident at a nuclear installation in the most emotive language, no matter how trivial it may be.

How Nuclear Power is Generated

Details of the technology are difficult to follow – as are the details of any modern technology – but in principle the processes which lead to the generation of electrical power in nuclear power stations are very simple. The operation begins with uranium. This is a metal that occurs naturally in most rocks, although some contain more than others, and in sea water. Uranium is all around us all the time and has been since the Earth first formed. It is an entirely natural substance. Uranium (chemical symbol U) occurs in three forms (isotopes) known as U-234, U-235, and U-238 and when the metal is extracted from its ore, uraninite, these three are always found in the proportions (rounded here to two decimal places so they do not add to precisely 100) 99.28 per cent U-238, 0.71 per cent U-235, and 0.006 per cent U-234. The number of each isotope represents the sum of the number of protons and neutrons in its atomic nucleus.

The nucleus of an uranium atom is unstable, so the metal itself is radioactive – it emits natural radiation in the form of alpha particles each consisting of two protons and two neutrons bound together, the loss of the two protons converting the uranium to thorium. The decay continues slowly, through a number of steps, until it reaches a stable condition beyond which no further decay occurs and the uranium has become lead.

About once for every million times an alpha particle is emitted, a U-238 nucleus divides into two smaller nuclei, with the emission of energy and several free particles, including neutrons. This is far too rare an event to be exploited, but U-235 is different. If a slow-moving neutron should collide with a U-235 nucleus it will be captured and held, making U-236. U-236 is extremely unstable and a collision from one more neutron will make it divide into two smaller nuclei, of yttrium-95 and iodine-139, with a release of energy, and then the yttrium and iodine also break down, releasing more neutrons and more energy. If sufficient U-235 can be concentrated in one critical mass, the neutrons that are released will have an excellent chance of colliding with another nucleus and so a self-sustaining chain reaction will occur.

U-235 is the fuel used in nuclear power stations and one tonne of it releases about as much energy as three million tonnes of coal. It is not used in the form of pure U-235 but, depending on the reactor design, usually as uranium oxide in which the uranium may be natural uranium or enriched uranium which has been processed to increase the proportion of U-235 to about 3 per cent. The fuel is sealed in metal tubes, called rods.

The neutrons released by the decay of U-235 travel fast. Fast neutrons are absorbed by the much more abundant U-238 and it is only the slow-moving neutrons that can be captured by U-235 and so sustain the reaction. The number of slow-moving neutrons can be increased by slowing down the fast-moving neutrons. To achieve this the fuel rods are inserted into a substance that has the property of slowing down neutrons that pass through it. Graphite (a natural form of pure carbon) will do this, and so will heavy water (deuterium oxide), or ordinary (light) water held under pressure. This substance is called the moderator. Rods of yet another substance are used to control the rate of the reaction. This substance, often boron, absorbs some of the neutrons that sustain the reaction, so lowering the rods into the moderator, among the fuel rods, makes the reaction proceed more slowly, and raising them accelerates it.

The moderator, fuel rods and control rods form the core of the reactor. The energy they produce is in the form of heat and this is carried away by a fluid – the primary coolant – which may be water or carbon dioxide. Should the primary cooling system fail, creating a risk that the core may overheat, a reactor is equipped with at least one, and usually more than one, emergency core-cooling system and, of course, the control rods can be dropped into the core rapidly, to damp down the reaction.

The heat carried away from the core is used to raise steam, the steam drives turbine generators, and that, in essence, is how a nuclear reactor works. It is no less natural than the burning of coal, oil, or gas. Indeed, there has been at least one entirely natural nuclear reactor.

The remains of this reactor were found some years ago at a place called Oklo, in Gabon. Some two billion years ago, geological processes concentrated enough uranium at Oklo to make a critical mass and allow a chain reaction to begin and the natural reactor ran until it exhausted its fuel. As a matter of interest and relevance the Oklo site has caused no pollution whatever of the local environment.

For some years now the world price for uranium has been fairly low but it cannot remain so for ever. When it rises a way will have to be found to use the metal more efficiently. This means using more than the 0.71 per cent that is U-235. Such a way exists. It employs fast breeder reactors. There is an experimental fast breeder in Britain, at Dounreay in Caithness, generating 250 megawatts, in France, the Superphénix demonstration plant generates 1.2 gigawatts, and a demonstration plant is being built in Japan. A fast breeder reactor can obtain 50 to 60 times more energy from uranium than can a conventional reactor.

The 'fast' refers to the fact that the reactor uses fast-moving neutrons and the 'breeder' to the fact that apart from generating electricity the reactor produces more fuel, in the form of plutonium (chemical symbol Pu) than it consumes. In most designs the reactor is cooled by molten sodium. This sounds hazardous, for sodium is highly reactive, but it is contained within a sealed system and in practice a breeder reactor is no more dangerous to operate than a conventional one.

A fast breeder reactor uses plutonium as a fuel, although it needs only an initial charge after which it produces its own, and it has no moderator. Instead, the core is surrounded by a 'blanket' of uranium. This can consist of spent fuel from conventional reactors in which the U-235 has been depleted. Neutrons from the decay of Pu-239 in the reactor core bombard the blanket, converting U-238 into more Pu-239. This plutonium can then fuel other reactors. In fact, the ordinary operation of a conventional reactor also produces plutonium. Most of this is burned in the reactor itself but a small amount remains in spent reactor fuel, from which it can be extracted. Plutonium is used either as a complete substitute for U-235 or as plutonium dioxide mixed with uranium dioxide to make a fuel called 'mixed oxide fuel' or 'MOX'. MOX is being used at two French reactors and there are plans for it to be used in several other countries. Another isotope, Pu-238, decays more rapidly and is used to power small batteries, for example in heart pacemakers.

Risks of Nuclear Power Generation

Fast breeder reactors produce plutonium, a dreaded name. Plutonium has been called 'the most dangerous substance in the world', although several other substances have also been described this way. It is true that plutonium is poisonous if you swallow it, although it is less poisonous than lead arsenate, cyanide or strychnine. Workers are more likely to inhale than to swallow plutonium and it is also poisonous if inhaled, but no more so than several other metals, such as cadmium and mercury. Plutonium is radioactive, of course, but the alpha radiation it emits cannot penetrate human skin, although the radiation of respiratory tissues by inhaled plutonium dust is dangerous. For safety reasons it is usual to wear gloves to protect the skin when handling plutonium but it is not really very dangerous in itself, especially when in the form of a solid lump of the pure metal. It is very reactive, however, and becomes coated with a layer of oxide immediately on being exposed to the air, and when it is in the form of a fine powder it can burn spontaneously, like sodium, magnesium, or uranium. In practice, therefore, it is usually handled inside a special cabinet filled with an inert gas, and it rarely exists as the pure metal. The form most commonly used industrially is plutonium dioxide, which is a solid, very hard material with a high melting point (2390 degrees C, 4334 degrees F).

Plutonium has also been described as 'wholly man-made'. This is not strictly true either. It does occur in uranium-bearing rocks, but in extremely small amounts, and presumably more must have been produced at Oklo, although it will all have decayed and disappeared by now.

All radioactive elements decay – it is what makes them radioactive – and as they do so their radioactivity decreases. Each isotope of each element decays at a characteristic and extremely constant rate which is usually expressed as the half-life; the time it takes for the radioactivity of that isotope to be reduced by one half. Obviously, the decay is exponential – it proceeds like the compound interest on a bank overdraft, only in reverse. If we give the initial radioactivity a value of 100 (the units do not matter), after one half-life it will be 50, after two 25, after three 12.5, after four 6.25, and so on. Half-lives vary widely. That of Pu-239 is 24,400 years, and of U-235 710 million years, but I-131, the radioactive isotope of iodine that was released in the 1957 Windscale and 1986 Chernobyl accidents, has a half-life of 8 days, and Rn-222, the most stable (that is, long-lived) isotope of radon has a half-life of 3.8 days.

A radioisotope can produce an effect only in proportion

to its level of radioactivity. As it decays, therefore, those effects diminish. The effects also depend on the amount of the radioisotope present to produce them. This complicates the evaluation of biological effects because the fate of any substance once it enters a living organism varies according to the substance itself. Iodine, for example, is concentrated by the human body in the thyroid gland, where it is needed for a range of necessary functions. If you ingest the radioactive isotope of iodine, therefore, it will be concentrated in your thyroid, which may be injured. In the event of an accidental discharge of radioactive iodine this risk diminishes with the half-life of iodine-131, so that after a couple of weeks or so it has disappeared, because it is then impossible for the body to concentrate a dose large enough to be harmful. When I-131 is released, therefore, strict measures are taken to limit human exposure to it and the acceptable limit of exposure is set very low. The acceptable limit of exposure to caesium-137, the other radioisotope most likely to be released in an accident, is much higher despite the fact that its half-life is 33 years. Caesium is not concentrated in any particular organ and so, being spread evenly throughout the body, much more of it is needed for the dose to be large enough to cause any harm.

After it has been operating for a while the whole core of any reactor becomes intensely radioactive, so it and its products must be isolated from the world outside. The reactor housing itself is contained within a shell designed to withstand any conceivable shock, so should the reactor fail, or even explode, its radioactive contents will not escape. The steam turbines need water, which must be taken into the plant and discharged again, but apart from the water itself in normal operation reactors emit virtually nothing. Indeed, coal-fired power stations emit up to 12 times more radiation than do nuclear power stations, most of the radioactive material consisting of potassium-40 which occurs naturally in coal. They also emit dust that contains cadmium, a metal as poisonous as plutonium but that is not radioactive and therefore never decays. If coal-fired power stations had to comply with the pollution limits imposed on nuclear stations they would be closed down at once.

No one disputes that exposure to high levels of ionizing radiation, and especially to electromagnetic radiation in the form of gamma and X rays, is lethal, and at very high doses gamma and X rays cause injury by direct burning - like sunburn. At somewhat lower doses people become ill, the severity of the illness depending on the extent of their exposure. They may make a complete recovery, but in some cases the damage caused by the radiation may lead to cancer later in life.

Ionizing radiation is so called because it penetrates tissue with sufficient energy to strip electrons away from their atoms, so ionizing the atoms. Once ionized, atoms become very reactive and it is the chemical reactions and their products that cause the harm by damaging cells. Cells that are only slightly damaged may recover. In the great majority of cases those that are more seriously damaged die or are destroyed by the body's own defence mechanisms and are removed. In some cases, however, the nucleic acids comprising the genes may be modified in such a way as to cause the affected cell to start proliferating rapidly. If such a cell is not destroyed by the body itself - and most are - the result can be cancer.

Ionizing radiation should be capable of producing genetic mutations that can be inherited but this has never been observed in humans. If it were, the victims themselves would almost certainly be incapable of reproducing because although mutations are common events, the vast majority of them are harmful and so eliminated - the genes, not necessarily the individuals bearing them. It is not true, therefore, that radiation can lead to the emergence of mutants - a race of hideously deformed people, or rats the size of cows. It can damage embryos and foetuses, leading to birth deformities, but this is not the same thing.

All of us are exposed to 'background' radiation throughout our lives. Of all the radiation we receive, 78 per cent is natural background radiation, the remainder artificial. These totals are made up of:

	%
Natural background:	
radon	33
gamma (from space and from rocks in the ground)	16
internal (from food and from our	

Natural background:	%
own bodies, as potassium-40 and carbon 14)	16
cosmic	13
Total natural	78
Artificial:	
medical (e.g. X rays)	20.7
occupational	0.4
fallout	0.4
miscellaneous	0.4
discharges from industry	0.1
Total artificial	22

This shows that the proportion of our total radiation exposure that is attributable to industrial releases is very small indeed, but it does not mean it is necessarily safe. Because a large dose is extremely dangerous and a smaller dose proportionately less dangerous, for safety reasons it has to be assumed that there is no such thing as a dose so small as to be harmless. This assumption leads to very stringent safety precautions but it is almost certainly incorrect. We have evolved in a radioactive environment and figures for background exposure are averages. Locally they can vary quite widely and some people are exposed naturally to much more radiation than others.

In recent years there has been great concern over the amount of radon entering homes, especially in those areas where the rocks are particularly rich in radium-bearing minerals. Radon is a radioactive gas produced naturally by the decay of radium-226 (half-life 1,602 years), a metal that occurs in uranium ores, and although radon is not particularly dangerous in itself some of its decay products, called radon daughters, are very short-lived but can adhere to surfaces so they are believed to be harmful if inhaled. Yet people have lived in such areas for thousands of years without coming to much harm and studies of cancer rates in high-radon areas in several countries show that people in those areas are no more likely to suffer from cancer than are people anywhere else. This is not surprising, for our evolution in a radioactive environment should have given us means of self-protection against ordinary radiation levels with an additional safety margin.

Indeed, there is a growing body of evidence to suggest that radiation doses just a little larger than the average natural background are actually beneficial. They stimulate the body's own defence mechanisms in much the same way as exposure to certain infections can bestow immunity against future infections by similar organisms.

The Sellafield 'Clusters'

Part of this evidence is contained in the report, published in 1987, of what is probably the most extensive and most detailed survey ever conducted of the health of people living close to nuclear installations. It was conducted in Britain, on behalf of the Office of Population Censuses and Surveys, covering the period from 1959 to 1980, and was started because of the concern about unusually large numbers of cases of childhood leukaemia that had been reported in the vicinity of Sellafield in Cumbria.

The Office of Population Censuses and Surveys is a British government institution. Some people have a distrust of government reports, suspecting they may be doctored to support a political view. In the case of many reports, especially those concerning economic and social matters, I am inclined to share this distrust and prefer to weigh the reports against those by other institutions.

In this case, however, there is good reason to be confident. The report deals with a matter susceptible to scientific evaluation and it is full and complete. By this I mean it is not merely a summary of findings and recommendations but a complete account of the way those findings were reached. It includes the data and details of the statistical methods involved. This means the interpretation, and therefore the findings, can be checked by others, and in fact they have been checked by a team drawn from some of the most experienced epidemiologists and statisticians, whose commentary on the report was published in a scientific journal. This fact also has implications. No such commentary, no matter how eminent and trustworthy its authors, can be published in scientific literature until it has been read, checked and approved by referees, other scientists, unconnected with the particular work or its authors, who are appointed by the editor.

It may be that the government, any government, might prefer a scientific study to arrive at a certain conclusion. There are such things as vested interests! In this case, however, it is difficult to see how the government could have succeeded in such a deception. Were the sampling inadequate, the study poorly conceived or conducted, or the findings not supported by the data, the scientific critics would have said so. What is more, the findings have received other, independent support recently.

The study dealt with all nuclear installations, not only power stations, and with the health of people living within a ten mile radius, and it dealt with all forms of cancer, not just leukaemias. The people in the study were divided into three age groups, 0–24, 25–74, and over 75, and were matched to control groups. The control populations were picked carefully to match the study groups in terms of total population size, proportion of urban and rural environments, and, as far as possible, social structure. The study found that the incidence of cancer in the study groups was lower than that in the control groups. This remained true even when the only cancers included were those particularly associated with exposure to radiation. In their commentary on the report in *Nature* (8 October, 1987) David Forman, Paula Cook-Mozaffari, Sarah Darby, Gwyneth Davey, Irene Stratton, Richard Doll, and Malcolm Pike said: 'Although many of these results will be chance findings as so many comparisons were made, the tendency for cancer mortality to be lower near nuclear installations than in the control Local Authority Areas is too strong to be explained solely on these grounds.'

Positive correlations were found, however, in the case of lymphoid leukaemia and all brain cancers among people aged 0–24 living within eight miles of installations built before 1955. The 'leukaemia cluster' around Sellafield was confirmed. The authors had dismissed the possibility that proximity to a nuclear installation could protect against cancer, concluding instead that the most likely explanation should be attributed to unintentional differences between the study groups and control groups. This led them to account for the positive correlations in the same way. The amount of radiation to which people in the study group were exposed was measured carefully and both the study and the authors of the commentary on it rejected the idea that the cancers could be related to radiation exposure.

Since 1987 new explanations have been proposed for the clusters. Many leukaemias are caused in the first place by viral infections. When a large industrial installation is built in a remote rural area the rural community has to absorb a large influx of people from other parts of the country. It is possible, therefore, that the incomers bring with them infections to which the host community has not previously been exposed and to which its members have acquired no immunity. This kind of thing has happened many times when colonists have settled a new area and mixed with a long-established and formerly isolated local population. If this explanation is correct, the clusters should not be confined to nuclear installations because these are not the only industrial complexes to have been built in rural areas; and they are not. Leukaemia clusters similar to the one around Sellafield have been found in other places, far from any nuclear installation. One was found in Gateshead, for example, and there are others in New Zealand.

In November 1989, a further report by Cook-Mozaffari, Darby, and Doll appeared in *The Lancet*. These scientists, financed by the Imperial Cancer Research Fund, compared mortality statistics from the regions close to two British nuclear power stations with those from six areas in which the Central Electricity Generating Board had identified possible sites for future nuclear power stations. They found the incidence of leukaemia and Hodgkin's disease among young people was raised but strikingly similar in both groups. This did not support the viral theory, because there had been no influx of people at the potential nuclear sites, but it did add to the mounting body of evidence against the idea that the cancers are caused by radiation, although debate about the hotspots at Sellafield and Dounreay is still open.

Whatever the cause, the seriousness of leukaemia clusters should be kept in proportion. If it is possible to prevent any deaths, of course we should do so, but we have to establish priorities. Something that kills ten people should be dealt with before we turn our attention to something that kills five. In Britain, leukaemias of all kinds cause 0.5 per cent of all deaths. Road accidents cause 1.2 per cent.

If the clusters can be explained without reference to radiation exposure – as they must be if those found far from nuclear installations are to be included in the explanation – we are left with the apparently curious finding that, in general, living close to a nuclear installation affords some protection against cancers of all kinds. It does look very much as though low-level radiation exposure can be beneficial.

Reactor Accidents

We can be sure, therefore, that nuclear reactors are perfectly safe and harmless during ordinary operation. In the event of a serious accident, however, large amounts of radioactive substances may be released.

This danger is quite real and has always been recognized. Like any responsible industry the nuclear industry learns by experience and each time a reactor component fails steps are taken to ensure that this particular fault does not recur. I have already mentioned the level of security incorporated in the building itself. You can measure our experience of any technology by adding together the number of plants using it and the number of years they have been in operation. This gives us something like 2,000 reactor-years of experience with the nuclear industry, during which most of the things that can go wrong have gone wrong.

All the same, accidents do occur and there will be accidents in the future. So far, there have been three incidents that have caused serious public alarm. Again, some people suspect the nuclear industry, until recently not renowned for the liberality with which it informed the public of its activities, may have concealed other incidents. In western Europe and the United States this is extremely unlikely. Incidents within nuclear plants could have been hidden, of course, so there may well have been accidents inside reactor installations of which we know nothing, but the release of radioactive substances to the environment outside a plant cannot be concealed. There are simply too many people monitoring and no way the discovery of elevated radiation levels could be kept out of the press.

The first of these accidents occurred at what was then

called Windscale (now Sellafield) in 1957. In that early design of reactor the coolant was air, drawn through the core and finally released into the atmosphere. Fuel rods overheated and caught fire leading to the release of the radioactive isotope iodine-131, which passed through the filters fairly easily because it existed as a gas, and small amounts of caesium and strontium. No one was injured but the iodine-131 contaminated pastures downwind and milk had to be destroyed. This reactor design is no longer used and so an accident of the Windscale type cannot happen again.

The second occurred on March 28, 1979, at Three Mile Island in Pennsylvania. It was very costly but so little radioactive material was released that no one suffered any injury.

The third, much more serious incident occurred at Chernobyl in the Ukraine on April 25, 1986. It did release a large amount of radioactive material which contaminated much of Europe to the north-west of the site.

The Chernobyl accident was peculiar to the type of reactor in which it occurred. This particular design uses graphite as a moderator, its fuel is contained in rods made from zirconium, and it is cooled by water carried in pipes through the core. At high temperatures and pressures water will react with both graphite and zirconium, releasing hydrogen. To reduce the risk of this the interior of the core housing contained an atmosphere of nitrogen, which will not support combustion. However, the housing itself was not contained within a secure shell, as are the housings of all Western reactors.

The reactor overheated and water was released, reacting with the graphite and zirconium and releasing hydrogen. An explosion breached the structure housing the reactor and allowed air to enter, and as air mixed with the hydrogen this caused a second, larger explosion. No other reactor design uses both graphite and water, so this reaction cannot occur except in Soviet reactors of the Chernobyl type. Nor is it possible in any other reactor system - or probably again in one using the Chernobyl system - for operators to shut down the emergency controls, which is how at Chernobyl a malfunction in the course of an experiment was able to develop into an uncontrollable catastrophe. When details of what happened were released, the magazine *New*

Scientist commented that if the operators tried to do to any Western reactor what they had done to the Chernobyl reactor not only would the computers governing the reactor refuse to accept their instructions, they would probably telephone the police!

Not only was this the worst accident ever to affect a nuclear installation, it was also the worst that could possibly happen. The reactor was destroyed completely and the explosions and subsequent fire went on releasing radioactive substances for some time before the building could be sealed. So how serious was it?

Undoubtedly the problems in the USSR were made worse by what can only be called bureaucractic incompetence. There was an unacceptably long delay before officials in Moscow would believe the reports they were receiving, and then there was confusion in the affected areas - confusion continuing to this day. At first people were advised to remain indoors. This was sensible as the cloud of contaminated material passed overhead - they were much safer indoors than they would have been in the open. Later, after the worst affected region had been evacuated, arguments began about the quality of the radiation monitoring, and much later still there were complaints that information had been suppressed and some people had been exposed to much higher doses than had been reported. There were stories of illness in humans and in livestock, and other, more reliable reports of changes affecting the leaves of trees - which generally seem to have been larger than usual. At a time of political turmoil in the USSR, when local people are demanding greater autonomy and increased economic support from Moscow, it is difficult to evaluate these stories. Some are clearly exaggerated, others may be true. Only time will tell.

In western Europe the effect was minimal but it did not seem so at the time. Earlier that year European governments had agreed new safety levels for human exposure to radiation. Their immediate reaction to Chernobyl was to reduce those levels arbitrarily to between one-tenth and one-hundredth of the values they had agreed only a few months previously. This was doubtless meant to reassure the public but it had precisely the opposite effect because it rendered 'dangerously radioactive' materials and foodstuffs that would have been considered harmless at the earlier, and very conservative, safety levels. The most serious harm occurred when herds of reindeer were slaughtered because meat from them was considered to be unsafe. In fact the meat would have harmed no one, but the Laplanders who relied on the reindeer were bureaucratically deprived of their means of livelihood. Much the same happened, but on a smaller scale, to sheep farmers in parts of Britain because of contamination of pastures by caesium-137. Restrictions on the sale of sheep from these areas continued for longer than had been anticipated because the caesium was not lost from the soil so rapidly as expected. The sheep were the beneficiaries, of course. Their health was not harmed and they were saved from slaughter!

According to the best information available at present, the immediate death toll at Chernobyl was 2 people killed in the explosion and a further 29 who died later, some from burns and other 'conventional' injuries, others from radiation sickness. More than 200 people suffered from radiation sickness but recovered, although some of them may develop cancer in years to come. The World Health Organization has estimated that as a result of the Chernobyl accident about 7,000 people may die prematurely from cancer over a period of 50 years, most of them citizens of the Ukraine and adjacent Byelorussia. This is a large number, but against the background of the 100,000 cancer deaths from other causes that will occur each year among the same population it will not be detectable statistically.

Appalling though they are, we should keep these figures in perspective. The industrial accident at Bhopal, in a factory making pesticides, killed more than 2,000 people, so serious as it was the Chernobyl accident is by no means the most devastating industrial accident in history. Nor does it make the nuclear power industry inherently more dangerous than its rivals. The World Health Organization has also estimated the number of deaths caused each year by diseases attributable to the burning of coal. For Britain the total comes to about 1,700. It means that coal-burning kills more people in Britain alone in a little over four years than the Chernobyl accident will kill in the entire world in 50 years.

Radioactive Waste

The final fear people have concerns the disposal of radioactive wastes, the problem to which anti-nuclear campaigners claim there is no solution, so that slowly, over the centuries, nuclear wastes will accumulate until they poison the whole world.

In fact the problem does not exist. The safe way to dispose of these hazardous wastes finally and permanently is well known, every aspect of it has been studied in great detail, and it is in use in several countries.

According to the intensity of the radiation they emit, radioactive wastes are classified as high-, medium-, and low-level. Low-level wastes consist of such things as discarded gloves and other clothing, laboratory glassware, and equipment, much of it from hospitals and research establishments. The regulations that define it as radioactive and therefore requiring special methods of disposal are based on radiation limits so low that many everyday substances, such as instant coffee, would fall within them were they to be produced by nuclear establishments. Those low-level wastes that are gases or liquids are stored until their radioactivity has fallen sufficiently for them to be released into the air or water. Solids are usually disposed of by burial, perfectly safely. High-level waste consists of spent reactor fuel and the wastes that remain after spent fuel has been reprocessed to extract the remaining U-235 and Pu-239 from it. It is in liquid form and is hot, intensely radioactive, and very dangerous indeed.

The waste is sealed in double-walled steel containers encased in concrete and cooled by agitating their contents and circulating water around them. In Britain no high-level waste has yet passed beyond this stage and the waste accumulated over nearly half a century is stored in tanks at Sellafield, from which there has never been any leakage. By the time it is processed in preparation for final disposal it will have decayed into medium-level waste.

The processing will begin by converting it into a particularly tough borosilicate glass. The cylindrical blocks of glass will be sealed in steel tanks and placed in water to cool. After about ten years they will have lost much of their heat and will be taken from the water and encased, probably in a thick jacket of iron. Then they will be taken to a store where air can flow over them to cool them further.

They will spend 50 to 70 years cooling and then they will be taken to their final disposal site, below ground or below the sea bed. The danger is that the containers may leak and their radioactive contents may contaminate ground water. Such contamination can occur only if the casings fail and the leaked material comes into contact with flowing water. The casings cannot be broken but they can corrode. Iron corrodes at a known rate, so if iron is used the thickness of the casing determines how long it will last under the worst conditions possible. Should the casing corrode so badly that holes develop in it the glass blocks will be exposed to whatever corroded the casing but their design ensures it would take up to 1,000 years for them to become so damaged as to allow radioactive substances to dissolve out of the glass. As an additional precaution, however, disposal sites will be chosen because they are dry, geologically stable, and either tend to seal themselves if ruptured or are able to absorb radiation. Salt and some clays are sufficiently plastic to reseal themselves should anything make holes in them.

The wastes must remain in their final disposal site for between 500 and 1,000 years depending on the particular chemical composition of the waste, by the end of which time they will be cool and harmless unless swallowed or inhaled. They will be no more radioactive than common uranium-bearing rocks. Quite apart from the security provided by the storage site itself, the processing of the wastes is designed to render them harmless for longer than the period for which they must be isolated. The method is based on the 'belt and braces' principle and allows a very large margin for safety.

All in all, the evidence leaves little room for doubt that nuclear power stations cause, and are capable of causing, less harm to the environment than those burning carbon-based fuels and less risk to human health than those burning coal. For health and environmental reasons, therefore, as existing coal-fired stations reach the end of their commercial lives it would be desirable to replace them with nuclear stations. This does not mean any overall expansion in generating capacity, or over-hasty expansion of the nuclear industry, only a gradual replacement in a carefully planned manner.

The Nuclear Future

In Britain, the possibilities for reducing the emission of carbon dioxide may have been seriously impaired owing to the determination of the Government to privatize the electricity supply industry. In November 1989, nuclear power stations were withdrawn from the privatization and it seemed that the nuclear industry might be allowed to contract, eventually perhaps to extinction. The decision resulted mainly from calculations of the cost of decommissioning nuclear plants at the end of their working lives when those costs had to be borne not by the public sector, but by private investors, who would not invest unless they were assured of a reasonable profit on their investment. Thus the decommissioning cost had not simply to be met, it had to be included in the overall cost of running the industry and made to be profitable. Not surprisingly, this raised the cost to a level markedly higher than that of generating electricity by the burning of coal, which is subject to much laxer environmental controls and, of course, does not experience decommissioning costs, only demolition costs. The anti-nuclear movement claimed a victory, but this may have been premature, for it was not certain the end of the nuclear industry had been ordained – and these costings and their effect on nuclear power apply only in Britain.

The demise of the nuclear industry would have consequences whose seriousness depends on your point of view. Were the industry to disappear altogether it is not only the capacity to generate power in this way that might be lost. Scientists and technologists upon whom we rely for advances in many aspects of nuclear technology might leave Britain. Were we then to change our minds, and decide that after all nuclear power does have a useful part to play in our economy, the knowledge we need, and probably the equipment as well, would have to be imported from those countries (the majority) in which the nuclear industry continues to thrive. You may or may not think this important.

The environmental consequences are no easier to predict. The course Britain is most likely to pursue will begin by increasing reliance on methane. This will lead to an increase in our emissions of carbon dioxide. When methane becomes scarce and therefore too expensive to burn, in a matter of a few decades from now, coal will probably be used instead, increasing carbon dioxide emissions still more and raising (because of increased demand) the price of coal and the cost of electricity.

This matters only if you suppose it desirable to reduce emissions of airborne pollutants in general and of greenhouse gases in particular. The other alternatives include increasing British imports of French (nuclear) electricity, (Britain is already France's best customer), or simply allowing the supply to fall below demand, in which case there will be power failures - brown-outs and black-outs.

British energy policy is in a state of some confusion. People have suggested rationing the use of carbon-based fuels by taxing them. It seems paradoxical, therefore, to abandon nuclear power because it is more costly than fossil-fuel-derived power.

Eventually, of course, the ordinary process of development renders all technologies obsolete. Nuclear reactor technology is no exception and, indeed, its replacement is already on the horizon. Present reactors are based on the principle of nuclear fission, the division of an atomic nucleus into two or more smaller nuclei with a release of energy. An alternative approach to nuclear power is based on fusion, the forced merger of two small nuclei.

In a fusion reactor the fuel may be deuterium, tritium, or lithium. Deuterium (D) and tritium (T) are both isotopes of hydrogen; lithium (Li) is a light metal. At very high temperatures and pressures various combinations of these nuclei - D-D, D-T, D-Li - can be made to fuse with a release of energy. As with fission reactors, the principle is simple but the technology is very complicated, the main difficulty arising from the fact that fusion can be made to occur only at temperatures so high no solid material can contain them without vaporizing. Stripped of their electrons, however, atomic nuclei carry a positive electrical charge which means they are affected by magnetic fields. In most experimental designs for fusion reactors the reaction takes place inside a strong magnetic field - a 'magnetic bottle'.

Research into fusion reactors is advanced and very promising but it will be some years yet before a commercial demonstration reactor can be built, and many years after that before fusion power starts to contribute significantly to our electricity supply. Fusion power is not likely to become commercially important before the second half of the next century. The investment is high, commercial fusion reactors will be very large and very expensive, but the advantages will be considerable. Of the hydrogen in ordinary water about 0.015 per cent is deuterium and it can be extracted fairly easily. Tritium is also present in water, but at a concentration too low to make it extractable from that source. Lithium, however, obtained from ores like other metals, can be transformed into tritium. If a deuterium-deuterium reactor can be built its fuel will be extracted from ordinary water. Even without that easy availability of an inexhaustible fuel the potential power output from a fusion reactor is very large indeed and the reactor itself is inherently safe. It is so difficult to make fusion take place at all that should any component fail the reactor will simply cease to work.

This apparent defence of nuclear power is necessary only to calm fears some of which are spontaneous but most of which result from deliberately alarmist campaigning. If the choice facing us is a free one, in the sense that it makes no real difference which decision we make, then we might phase out our existing reactors with no more than a tinge of regret for having done so without any real justification.

We could not switch to wind, wave, solar or tidal power because they are incapable of making more than a very small contribution to our need. Supplies of petroleum are limited so it would be unwise to place too much reliance on them. We could make greater use of methane because although natural gas supplies are also limited it is possible to make methane from coal – the resulting gas is not the poisonous 'town gas' we used years ago, whose 'active ingredient' was carbon monoxide. Most obviously, we could rely more heavily on coal. That would provide employment, albeit hazardous employment, for miners, with the additional advantage that Britain still has sufficient supplies of its own to last many years without having to turn to imports, although imported coal, from mainly shallow pits or open-cast sites, is often cheaper than our own deep-mined coal. Advances in furnace design may even allow us to burn coal more cleanly.

The choice is not a free one, however. Carbon dioxide is accumulating in the atmosphere and the burning of carbon-based fuels is a major source. If we seek to reduce the rate of its accumulation we must burn less oil, gas and coal and once we accept the inevitability, and desirability, of greater industrialization in what is now the Third World that need becomes still more imperative. Far from being free, our choice disappears altogether. We have no choice, for nuclear power is the only means we have to provide ourselves with the electricity we need. That being so we should try to understand what it is, how it works, and to distinguish between genuine risks and those we may perceive but that we exaggerate.

CONCLUSION

Crossroads and Signposts

For the last 20 years people have been telling us that we stand at the crossroads. Perhaps we have always been at the crossroads, immobilized by a confusion engendered by the signposts, which keep changing. Indeed, the very nature of the crossroads themselves keeps changing. Sometimes the destinations offered are political, sometimes economic, at other times social, or medical, and very often they are environmental. The only certainty is that a dismal fate awaits those who make the wrong decision. I suspect that in the course of history we have been at this place many times before. The view from here has a haunting familiarity.

Many people insist that all will be well if we turn back, retracing our steps to a better time and more congenial landscape but when we peer in the direction of the past all that we see is a firmly blocked road. It might be pleasant to travel in that direction but the way is barred. Few of our advisers recommend proceeding straight ahead, which, so far as most of us can see, is the simple and obvious thing to do. You can see their point. It would hardly be a crossroads if we could safely continue on our present track, ignoring what would then be irrelevant side-roads. So we are strongly advised to make a turning, with strident voices competing to draw us along the route each of them prefers.

Now the crossroad has changed yet again, and so has the map on which it appears. The map is now called 'climate' and, as usual, we are not to proceed forward and cannot proceed backward.

We should not be too harsh with the cartographers for they are reformers and no reform has ever been achieved without simplifying and over-dramatizing the issues. Reform implies a change of direction, which automatically precludes any continuation along our present path and we are more likely to hear and understand the reformer who points out clearly the direction in which we must turn and describes lucidly the destinations that lie at the end of each alternative road. This involves simplification, of course, the blanking out of much of the detail on the map, and it may also involve somewhat apocalyptic descriptions of the destinations we are to reject.

Perhaps, then, we should not be too harsh with the cartographers but we can be just a little bit harsh. Where the map is genuinely intricate and subtle the special pleading of reformers may merely compound our confusion and predictions of doom may inhibit reform rather than promoting it. If the only choice presented to us is so unpalatable we cannot bear to accept it, the possibility of doom at some time in the remote future may seem preferable and we can console ourselves with the one fact of which we can be truly sure: by the laws of physics and the cosmos, the future is unpredictable. For want of the option of a small change we reject all change and do nothing.

In this book I have tried to fill in some, but by no means all, of the detail that is omitted in some accounts of possible climate change. I have tried to provide a more complete map, if you like, because I believe that will make it easier for us to reach sensible decisions. Better informed judgements are usually sounder judgements.

The climate is the product of interactions between the air, oceans, and land masses, driven by energy from the Sun. The weather is the local manifestation of climate. The global climate is extremely complicated and there is much about it that we do not know. Even if we possessed complete knowledge it does not necessarily follow that we could make accurate predictions, for the reactions of which it consists may be highly sensitive to innumerable very small events, so that in unforeseeable ways apparently similar causes may result in widely different effects.

If the climate is changing on a global scale perhaps we do not need detailed forecasts of the consequences for every corner of the world. It may be enough to have a broad, coarse-grained picture to guide our response. Certainly this is the view of most people. I question it, however, just because of the possibility that the fine detail of pressures, temperatures, humidities, and winds at a very local level may exert a quite disproportionate influence on the broad pattern of climate.

As it is, the information on which predictions are made is quite general and the best computer models provide pictures that are very coarse-grained indeed, like maps that depict little more than smoothed-out coastlines and main roads. They are better than no maps at all, of course, but they must be interpreted with care. We must remember that even at their large scale the formation of clouds and their types can be described only in the most approximate way and that the influence of sea-surface temperatures is treated in even more general terms.

We do know that certain climatic changes are occurring. It is growing warmer, ice caps are retreating, at least in some places, and rainfall patterns are somewhat different in several parts of the world. Unfortunately, each such change taken separately, or all of them taken together, can be explained satisfactorily without invoking a major interference resulting from human activities. There has been freak weather, too, but sensational though it is when it occurs this is even less reliable as an indication of anything at all. There has always been freak weather.

It may strike you, therefore, that if a climatic crossroads exists we have not yet reached it and should make no decision at all until we have much more reliable information. This may be the case, but there is another possibility. It may be that the crossroads does exist, we have reached it and we are still moving forward and cannot stop, but for lack of information we are unable to make out what it says on the signposts. We should bear in mind that although the models may be crude they are tested by making them 'predict' past climates which can be checked against historical records, and they are not permitted to advance into the future until they have passed that test. We should also remember that the present rate of atmospheric accumulation of particular gases is a matter of direct measurement, not estimate or approximation, and whether particles and molecules reflect, absorb and re-radiate, or are transparent to radiation at each wavelength is accepted fact, and not in dispute.

The signpost seems to me to be inscribed with curious hieroglyphics from which I am able to extract several possible meanings. The first tells me that the climate will remain very much as it is now no matter what we do or fail to do. The second tells me that the present warming is no more than the ending of the Little Ice Age, that it will soon end, and that the underlying trend is toward a cooling and eventually to another full ice age. The third tells me that human activities have indeed triggered a warming that will become evident within the next decade or two, but that the natural response, by the planet itself as it were, will reverse it rapidly so the climate will start to warm but then cool rapidly, though perhaps not everywhere and not to the extent of an ice age. The fourth tells me that the warming of the climate will soon accelerate to bring warmer conditions to most places and drier conditions to the interiors of all continents. The fifth tells me that the warming will bring warmer and wetter conditions everywhere. Perhaps I should point out here that the rules of the game forbid the tearing down and burning of the signpost itself!

Many people fear that any rapid climatic change is bound to cause great disruption and there is little doubt that a change for the worse would necessitate major human migrations. If the grain-growing region of North America were to shift northward, for example, Canadians would benefit but what would happen to people living in and dependent on the southern part of the US grain belt? Not all the predictions are of changes for the worse, however. A warmer, wetter world, in which deserts retreated and eventually disappeared, would be a distinct improvement especially at a time when each year brings us more mouths that must be fed.

So far as the natural flora and fauna are concerned the warnings are less convincing, for rapid climatic change has taken place in the past and they have survived. We are often told of the 10,000 years that have elapsed since the official ending of the last ice age, as though those years had seen a very gradual warming of the climate. It is not what happened, or what happens. The climate warmed in fits and starts, jumping several degrees in a few decades, then sticking or even growing colder, then warming again. Cooling happens in the same way and it is possible that the climate can switch within a century or so from mild conditions like those we enjoy today to a full ice age. Species have always adapted, though they might find it more difficult to adapt now in places where human land management has confined them to isolated 'oases' from which escape may be prevented by landowners protecting their domesticated plants and animals.

Amid so much contradiction it is not easy to see how we should respond or even whether we should respond at all – if conditions are going to improve we would be foolish to intervene. Yet perhaps we should be cautious, lest the more unpleasant predictions prove correct, and take what steps we can to reduce our emissions of the greenhouse gases everyone agrees are capable of altering the climate. Such caution allows us a compromise. We might take action whose outcome is beneficial for other reasons. If our reforms benefit us anyway, then they will not be pointless if their equally benign influence on the climate proves irrelevant.

There are some things we have already decided to do. We have agreed to phase out rapidly the use of CFCs in order to prevent the depletion of the ozone layer. CFCs are greenhouse gases and a thinning of the ozone layer would lead to some warming of the lower atmosphere, so by ceasing to use these chemicals we will influence the climate. There is a risk, however, that the chemicals introduced to replace CFCs will also be greenhouse gases and, of course, if it should transpire that we face a cooling rather than a warming we might wish to deplete the ozone layer deliberately.

Conservationists would welcome any measure that would prevent further clearance of tropical forests. This would also influence climate since forest clearance, especially when accompanied by the burning of vegetation, releases greenhouse gases. The forests are cleared partly through the greed of rich corporations in pursuit of large profits and partly through the desperation of poor people seeking land on which to grow food. If we can curb the excesses of greed it will be no bad thing, and if we can relieve poverty, allowing landless peasants access to better land that is known to be cultivable, that will be a very good thing. So saving the forests may also involve redressing economic injustice which will make the world a happier, better and probably safer place.

The release of methane by farming can be restricted only by reforms in livestock husbandry in temperate countries, for even if it were possible it would be quite wrong to restrict the cultivation of rice by people for whom it is a staple food. Such reforms, I suggest, might involve encouraging the raising of cattle only on grass. This would affect the mainly American feedlot system, but that system generates its own serious slurry disposal problems and often causes local pollution so there would be environmental gains as well as economic losses. Provided we made sure the decline in output was not made good from ranches on land carved out by clearing South American forests, the overall result might mean we ate less beef, but nutritionists would not object to that.

The biggest gains, however, will come from changes in the way we produce and use energy. If we can economize in the amount of fuel we burn we will benefit in several ways. The burning of coal is one of the world's major sources of pollution and causes much damage, not least to human health. If we burn less of it we will be healthier and the world will be cleaner. That is a real advance. Fuel of all kinds costs money, and so if we use less of it we will also be wealthier. That, too, will improve the quality of our lives.

The benefits deriving from some possible reforms are less direct. A change in emphasis from private to public transport systems might save fuel while leading to greater mobility for everyone, but most of all for those who do not own cars or, like many housewives, do not have easy access to them. It would also reduce air pollution and ease traffic congestion. Cars that use less fuel will save money for motorists, but cars and other items of everyday equipment such as washing machines and refrigerators that are built to last longer would also save energy in the production process, as well as materials. They might be more expensive to buy, but their greater durability might well make them cheaper in the end.

No matter how great may be the economies of which we are capable, we will continue to need a large and reliable supply of energy and we will have to consider alternatives to the burning of carbon-based fuels. We who live in advanced industrial countries have a special obligation here if citizens of the less developed countries are to increase their industrial output using the technologically less demanding carbon-based fuels while over the world as a whole we are to prevent further increases in emissions of carbon dioxide. Most environmentalists favour the so-called 'renewable' sources of energy, but each of these suffers from formidable environmental disadvantages, especially in crowded Britain, and according to the most optimistic forecasts all of them together can contribute only a very small proportion of the energy we will need.

Whether we like the idea or not, a reduction in carbon dioxide emissions requires us to increase our reliance on nuclear power. Because this conclusion seems unavoidable I have tried to explain what nuclear power is and how it works, and to put the perceived dangers into perspective. It may be that a more realistic appraisal of the relative merits and demerits of the alternative methods for generating electricity will lead us to the strong suspicion that the replacement of carbon-based fuels with uranium would in any case reduce pollution of the environment. If so, this is also a reform that can be justified on other than climatic grounds.

In times of uncertainty it is wise to remain optimistic but to prepare for the worst. Mild optimism may lead us to hope for no climatic change at all. A strongly optimistic outlook may allow us to hope for an improvement on a world-wide scale. Such hopes can be justified, but until the future reveals itself they are merely hopes and hopes they must remain, and it would be foolhardy to pretend otherwise. They are not the future for which prudence dictates that we should prepare. Rather we should plan for the possibility that the climate will deteriorate, becoming either colder, or warmer but in places drier.

So far as the climate is concerned, at this stage planning is as far as we are able to go. So far as the wider global environment is concerned we can go further, and perhaps we should; for regardless of what may happen climatically, there are appropriate reforms we might make whose outcome would be a cleaner, healthier, and more equitable world.

Index